SpringerBriefs in Fire

T0203095

Series Editor
James A. Milke

For further volumes:
http://www.springer.com/series/10476

Celina Mikolajczak · Michael Kahn
Kevin White · Richard Thomas Long

Lithium-Ion
Batteries Hazard
and Use Assessment

 Springer

Celina Mikolajczak
Exponent Failure Analysis
    Associates, Inc.
Menlo Park
CA 94025
USA

Kevin White
Exponent Failure Analysis
    Associates, Inc.
Menlo Park
CA 94025
USA

Michael Kahn
Exponent Failure Analysis
    Associates, Inc.
Menlo Park
CA 94025
USA

Richard Thomas Long
Exponent Failure Analysis
    Associates, Inc.
Menlo Park
CA 94025
USA

ISSN 2193-6595
ISBN 978-1-4614-3485-6
DOI 10.1007/978-1-4614-3486-3
Springer New York Heidelberg Dordrecht London

e-ISSN 2193-6609
e-ISBN 978-1-4614-3486-3

Library of Congress Control Number: 2012933841

Printed on acid-free paper

Springer is part of Springer Science+Business Media (www.springer.com)

Preface

Lithium-ion (Li-ion) has become the dominant rechargeable battery chemistry for consumer electronics devices and is poised to become commonplace for industrial, transportation, and power-storage applications. This chemistry is different from previously popular rechargeable battery chemistries (e.g., nickel metal hydride, nickel cadmium, and lead acid) in a number of ways. From a technological standpoint, because of high energy density, lithium-ion technology has enabled entire families of portable devices such as smart phones. From a safety and fire protection standpoint, a high energy density coupled with a flammable organic, rather than aqueous, electrolyte has created a number of new challenges with regard to the design of batteries containing lithium-ion cells, and with regard to the storage and handling of these batteries. Note that energy storage is an area of rapidly evolving technology. There are a number of efforts underway to commercialize cells with different chemistries than lithium-ion including rechargeable lithium metal cells, ultracapacitors, and fuel cells. It is beyond the scope of this document to describe the characteristics and hazards of all of these potential energy storage devices.

At the request of the Fire Protection Research Foundation (FPRF), Exponent assessed the potential fire hazards associated with lithium-ion batteries. This assessment was intended to be a first step in developing fire protection guidance for the bulk storage and distribution of lithium-ion batteries both alone and in manufactured products. This report contains seven chapters:

- Chapter 1—Provides a general introduction to lithium-ion cells (Fig. 1) and batteries (Fig. 2). It includes a basic description of how lithium-ion cells function and are typically constructed, how various lithium-ion cells are characterized (chemistry, form-factor, case material, size), and how cells are combined to form battery packs.
- Chapter 2—Provides a discussion of lithium-ion battery applications It includes a discussion of the variety of ways lithium-ion cells are currently implemented, including: medical devices, consumer electronics, automotive applications, aerospace applications, and stationary power applications.

Fig. 1 A selection of typical consumer electronics lithium-ion cells

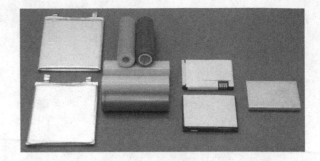

Fig. 2 A selection of typical consumer electronics lithium-ion battery packs

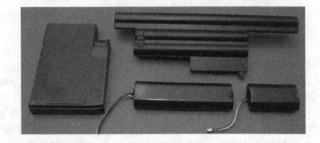

- Chapter 3—Provides a summary of applicable codes and standards. Particularly, the various transportation and safety standards that currently apply to lithium-ion cells and batteries as well as some of the standards that are available or being drafted specific to automotive applications of lithium-ion cells are discussed.
- Chapter 4—Discusses lithium-ion battery failure modes. It includes a discussion of various known lithium-ion failure modes and when during a cell or battery pack's life cycle they are most likely to occur (e.g., storage, transport prior to usage, early usage, after extended usage, during transport for disposal) as well as under what usage conditions a failure is likely to occur.
- Chapter 5—Discusses the typical life cycle of a lithium-ion cell or battery pack. It focuses on handling, transport, and storage procedures used at the various stages of battery life cycle from cell manufacture through cell recycling.
- Chapter 6—An assessment of the potential fire hazards associated with transport and storage of lithium-ion batteries.
- Chapter 7—Discusses gaps in data relevant to fire protection issues and testing approaches to address those gaps.

In general, this report focuses on aspects of lithium-ion cell and battery designs that are of particular significance to fire protection professionals.

Acknowledgments

The authors would like to thank the FPRF for giving Exponent the opportunity to complete this work. We would also like to thank a number of our colleagues at Exponent who provided assistance and advice, particularly: Hubert Biteau, Scott Dillon, Priya Gopalakrishnan, Troy Hayes, Quinn Horn, Mikhail Kislitsyn, Don Parker, Ryan Spray, and Ming Wu.

Contents

Acronyms and Abbreviations

18650	A common cylindrical cell form factor/designation
A	Ampere
Ah	Ampere-hour
ARC	Accelerating rate calorimetry
ATA	Air Transport Association
BMS	Battery management system
BMU	Battery management unit
BTU	British thermal unit
°C	Degrees Celsius
CAA	Civil Aviation Authority of the United Kingdom
CEI	Commission Electrotechnique Internationale
CEN	European Committee for Standardization
CID	Charge interrupt device
CFR	Code of Federal Regulations
CTIA	The Wireless Association
DEC	Diethyl carbonate
DMC	Dimethyl carbonate
DOT	United States Department of Transportation
DSC	Differential scanning calorimetry
EC	Ethylene carbonate
EV	Electric vehicle
°F	Degrees Fahrenheit
FAA	Federal Aviation Administration
FLA	Flooded lead acid
FPRF	Fire Protection Research Foundation
g	Gram
HEV	Hybrid electric vehicle
HF	Hydrofluoric acid
IATA	International Air Transport Association
ICAO	International Civil Aviation Association
IEC	International Electrotechnical Commission

IEEE	Institute of Electrical and Electronics Engineers
IMDG	International Maritime Organization
INERIS	L'Institut National de l'Environnement Industriel et des Risques
ISO	International Organization for Standardization
JEVA	Japan Electric Vehicle Association
kcal	Kilocalorie
kJ	Kilo-Joule
LFL	Lower flammability limit
m	Meter
ml	Milliliter
Mm	Millimeter
NFPA	National Fire Protection Association
NiCad	Nickel cadmium
NiMH	Nickel metal hydride
NRIFD	National Research Institute of Fire and Disaster (Japan)
PC	Propylene carbonate
PCB	Printed circuit board
PHEV	Plug-in hybrid electric vehicle
PRBA	The Rechargeable Battery Association
PTC	Polymeric thermal cutoff
RBRC	Rechargeable Battery Recycling Corporation
SAE	Society of Automotive Engineers
SEI	Solid electrolyte interphase
SOC	State-of-charge
TGA	Thermo gravimetric analysis
UFL	Upper flammability limit
UL	Underwriters Laboratories
ULD	Unit load device
UN	United Nations
V	Volts
VRLA	Valve regulated lead acid
W	Watt
Wh	Watt-hour

Chapter 1
Introduction to Lithium-Ion Cells and Batteries

The term lithium-ion (Li-ion) battery refers to an entire family of battery chemistries. It is beyond the scope of this report to describe all of the chemistries used in commercial lithium-ion batteries. In addition, it should be noted that lithium-ion battery chemistry is an active area of research and new materials are constantly being developed. This chapter provides an overview of the technology and focuses on the characteristics of lithium-ion batteries common to the majority of available batteries. Additional detailed information with regard to lithium-ion batteries is available in a number of references including *Linden's Handbook of Batteries*,[1] *Advances in Lithium-Ion Batteries* edited by Schalkwijk and Scrosati,[2] and a large volume of research publications and conference proceedings on the subject.

In the most basic sense, the term lithium-ion battery refers to a battery where the negative electrode (anode) and positive electrode (cathode) materials serve as a host for the lithium ion (Li+). Lithium ions move from the anode to the cathode during discharge and are intercalated into (inserted into voids in the crystallographic structure of) the cathode. The ions reverse direction during charging as

---

[1] *Linden's Handbook of Batteries*, 4th Edition, Thomas B. Reddy (ed), McGraw Hill, NY, 2011.
[2] *Advances in Lithium-Ion Batteries*, WA van Schalkwijk and B Scrosati (eds), Kluwer Academic/Plenum Publishers, NY, 2002.

C. Mikolajczak et al., *Lithium-Ion Batteries Hazard and Use Assessment*, SpringerBriefs in Fire, DOI: 10.1007/978-1-4614-3486-3_1,

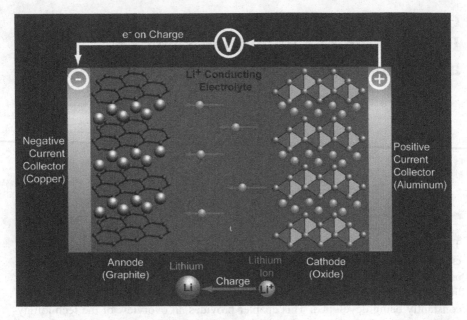

Fig. 1.1 Lithium-ion cell operation, during charging lithium ions intercalate into the anode, the reverse occurs during discharge

shown in Fig. 1.1. Since lithium ions are intercalated into host materials during charge or discharge, there is no free lithium metal within a lithium-ion cell,[3, 4] and thus, even if a cell does ignite due to external flame impingement, or an internal fault, metal fire suppression techniques are not appropriate for controlling the fire.

---

[3] Under certain abuse conditions, lithium metal in very small quantities can plate onto anode surfaces. However, this should not have any appreciable effect on the fire behavior of the cell.

[4] There has been some discussion about the possibility of "thermite-style" reactions occurring within cells (reaction of a metal oxide with aluminum, for example iron oxide with aluminum, the classic thermite reaction, or in the case of lithium-ion cells cobalt oxide with aluminum current collector). Even if thermodynamically favored (based on the heats of formation of the oxides), generally these types of reactions require intimate mixtures of fine powders of both species to occur. Thus, the potential for aluminum current collector to undergo a thermite-style reaction with a cathode material may be possible, but aluminum in bulk is difficult to ignite (Babrauskas V, *Ignition Handbook*, Society of Fire Protection Engineers, 2003, p. 870) and thus, the reaction may be kinetically hindered. Ignition temperatures of thermite style reactions are heavily dependent upon surface properties. Propagation of such reactions can also be heavily dependent upon mixture properties. To date, Exponent has not observed direct evidence of thermite style reactions within cells that have undergone thermal runaway reactions, nor is Exponent aware of any publically available research assessing the effect of such reactions on cell overall heat release rates. Nonetheless, even if a specific cell design is susceptible to a thermite reaction, that reaction will represent only a portion of the resulting fire, such that the use of metal fire suppression techniques will remain inappropriate.

Fig. 1.2 Example of a stacked prismatic cell design

Fig. 1.3 Base of a cylindrical lithium-ion cell showing wound structure (*top*). Cell being unwound revealing multiple layers: separator is white, aluminum current collector (part of cathode) appears shiny (*bottom*)

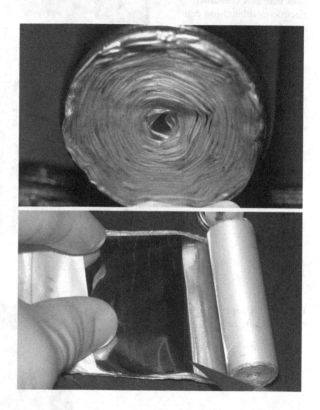

In a lithium-ion cell, alternating layers of anode and cathode are separated by a porous film (separator). An electrolyte composed of an organic solvent and dissolved lithium salt provides the media for lithium ion transport. A cell can be constructed by stacking alternating layers of electrodes (typical for high-rate capability prismatic cells) (Fig. 1.2), or by winding long strips of electrodes into a "jelly roll" configuration typical for cylindrical cells (Figs. 1.3 and 1.4). Electrode stacks or rolls can be inserted into hard cases that are sealed with gaskets (most commercial cylindrical cells) (Fig. 1.5), laser-welded hard cases (Fig. 1.6), or enclosed in foil pouches with heat-sealed seams (commonly referred to as

Fig. 1.4 Computed
tomography scan (CT scan)
of an 18650 cell showing
structure in cross section

Fig. 1.5 Examples of 18650
cylindrical cells (these are the
most common consumer
electronics lithium-ion cell
form factor)

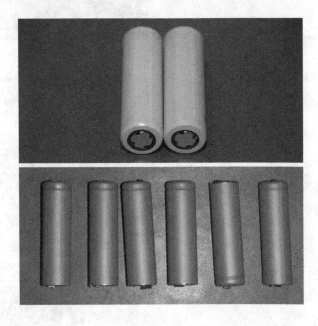

Fig. 1.6 Example of a hard
case prismatic cell

Fig. 1.7 Example of a soft-pouch polymer cell

lithium-ion polymer cells[5]) (Fig. 1.7). A variety of safety mechanisms might also be included in a cell mechanical design such as charge interrupt devices and positive temperature coefficient switches.[6,7]

An individual lithium-ion cell will have a safe[8] voltage range over which it can be cycled that will be determined by the specific cell chemistry. For most commercial lithium-ion cells, that voltage range is approximately 3.0 V (discharged, or 0% state-of-charge, SOC) to 4.2 V (fully charged, or 100% SOC). Because of a relatively flat discharge profile, the "nominal" voltage (voltage that the cell will exhibit through most of its discharge) of a typical lithium-ion cell is usually approximately 3.6–3.7 V. For most cells,[9] discharge below 3.0 V can cause degradation of electrodes and thus discharge below the manufacturer's low voltage specification is referred to as over-discharge. Repeated over-discharge can lead to cell failure and cell thermal runaway (discussed below). For most cells,[10] charging significantly above 4.2 V (e.g., to 5 V) can lead to rapid, exothermic degradation

---

[5] Note that the term "lithium polymer" has been previously used to describe lithium metal rechargeable cells that utilized a polymer-based electrolyte. The term lithium polymer is now used to describe a wide range of lithium-ion cells enclosed in soft pouches with electrolyte that may or may not be polymer based.

[6] For a more detailed discussion of lithium-ion cells see: Dahn J, Ehrlich GM, "Lithium-Ion Batteries," *Linden's Handbook of Batteries*, 4th Edition, TB Reddy (ed), McGraw Hill, NY, 2011.

[7] For a review of various safety mechanisms that can be applied to lithium-ion cells see: Balakrishnan PG, Ramesh R, Prem Kumar T, "Safety mechanisms in lithium-ion batteries," Journal of Power Source, 155 (2006), 401–414.

[8] A safe voltage range will be a range in which the cell electrodes will not rapidly degrade due to lithium plating, copper dissolution, or other undesirable reactions.

[9] Some specialty lithium-ion cells are available commercially that allow discharge to 0 V (e.g., see http://www.quallion.com/sub-mm-implantable.asp).

[10] Some commercially available lithium-ion cells can be charged to higher than 4.2 V; however, these are fairly rare.

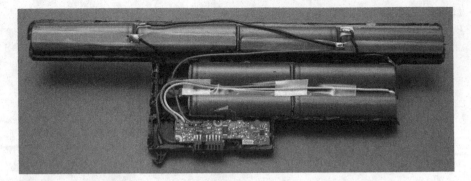

Fig. 1.8 An example of a battery pack that contains multiple cells (in *red* shrink-wrap) and a pack protection printed circuit board (PCB) (*green*)

Fig. 1.9 Schematic of cells connected in parallel

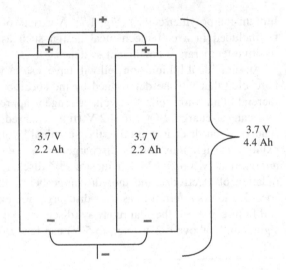

of the electrodes. Charging above the manufacturer's high voltage specification is referred to as overcharge. Since overcharging can lead to violent thermal runaway reactions (see footnote 1), a number of overcharge protection devices are either designed into cells or included in the electronics protection packages for lithium-ion battery packs.

A lithium-ion battery (or battery pack) is made from one or more individual cells packaged together with their associated protection electronics (Fig. 1.8). By connecting cells in parallel (Fig. 1.9), designers increase pack capacity. By connecting cells in series (Fig. 1.10), designers increase pack voltage. Thus, most battery packs will be labeled with a nominal voltage that can be used to infer the number of series elements and pack capacity in Ampere hours (Ah) or Watt hours (Wh) that will provide an indication of the capacity of each series element

Fig. 1.10 Schematic of cells connected in series

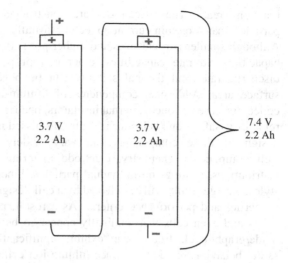

3.7 V
2.2 Ah

3.7 V
2.2 Ah

7.4 V
2.2 Ah

(size of individual cells or the number of cells connected in parallel). For example, a lithium-ion battery pack marked as 10.8 V nominal, 7.2 Ah can be assumed to contain three series elements (3 × 3.6 V = 10.8 V), with each series element containing 7.2-Ah capacity. Typical 18650-sized cylindrical cells (18650 cells are the consumer electronics workhorse cell—they are found in most multi-cell battery packs) at the time of this writing, have capacities that range from 2.2 to 2.9 Ah; thus, a notebook computer battery pack with a 7.2-Ah capacity label would likely include series elements containing three 2.4-Ah cells connected in parallel, and the entire battery pack contains nine cells in a 3s, 3p arrangement (i.e., 3 series elements containing 3 cells each in parallel).

For large format battery packs, cells may be connected together (in series and/ or in parallel) into modules. The modules may then be connected in series or in parallel to form full battery packs. Modules are used to facilitate readily changed configurations and easy replacement of faulty portions of large battery packs. Thus, large format battery pack architecture can be significantly more complex than small consumer electronics battery packs which typically contain series connected elements consisting of two or more parallel connected cells. Nonetheless, the simplified analysis method used above can still be applied to generally understand the total number of series elements within a battery pack, and the capacity of the parallel elements.

New UN regulations require that a battery pack be labeled in terms of Wh, which is battery pack capacity expressed in Ah multiplied by nominal voltage. Thus, a 7.2-Ah battery pack containing cells with nominal voltages of 3.6 V might be labeled a 25.9 Wh battery pack.

The four primary functional components of a practical lithium-ion cell are the negative electrode (anode), positive electrode (cathode), separator, and electrolyte. To increase the battery's storage capacity it is desirable for the anode and cathode materials to have large geometric electrode areas with high porosity to increase reaction area.[11] Thus, electrodes are constructed of pastes composed of fine particles coated on thin current collectors (usually thin copper or aluminum foils). Although smaller particle sizes and higher porosities will generally lead to higher capacities and rate capabilities, other cell properties such a cycle life, self-discharge rate, and thermal stability can be negatively affected by increased surface area. Additional components of lithium-ion cells such as the current collectors, case or pouch, internal insulators, headers, and vent ports also affect cell reliability, safety, and behavior in a fire (discussed in Chap. 4). The chemistry and design of these components can vary widely across multiple parameters. Cell components, chemistry, electrode materials, particle sizes, particle size distributions, coatings on individual particles, binder materials, cell construction styles, etc., generally will be selected by a cell designer to optimize a family of cell properties and performance criteria. As a result, no "standard" lithium-ion cell exists and even cells that nominally appear to be the same (e.g., lithium cobalt oxide/graphite electrodes) can exhibit significantly different performance and safety behavior. In addition, since lithium-ion cell chemistry is an area of active research, one can expect cell manufacturers will continue to change cell designs for the foreseeable future.

The market is currently dominated by lithium-ion cells that have similar designs: a negative electrode made from carbon/graphite coated onto a copper current collector, a metal oxide positive electrode coated onto an aluminum current collector, a polymeric separator, and an electrolyte composed of a lithium salt in an organic solvent.

Negative Electrode (Anode)

The lithium-ion cell negative electrode is composed of a lithium intercalation compound coated in a thin layer onto a metal current collector. The most common anode material is some form of carbon, usually graphite, in powder form, combined with binder material.[12] The nature of the carbon can vary considerably: in the source of the graphite (natural or synthetic), purity, particle size, particle size distribution, particle shapes, particle porosity, crystalline phase of carbon, degree of compaction, etc. Anodes composed of silicon, germanium, and Titanate ($Li_4Ti_5O_{12}$) materials have also been produced or tested, but at the time of this writing, non-graphitic anodes are rarely implemented.

---

[11] Brodd RJ, Tagawa K, "Lithium-Ion Cell Production Processes," *Advances in Lithium-Ion Batteries*, WA van Schalkwijk and B Scrosati (eds), Kluwer Academic/Plenum Publishers, NY, 2002.

[12] For a detailed discussion of carbon anode materials, see: Ogumi A, Inaba M, "Carbon Anodes," *Advances in Lithium-Ion Batteries*, WA van Schalkwijk and B Scrosati (eds), Kluwer Academic/Plenum Publishers, NY, 2002.

Thin, uniform coatings of active materials are required in lithium-ion cells that use organic electrolytes (at the time of this writing almost all commercially available cells). Thus, the negative electrode material mixing and coating process is often proprietary as variations in processing parameters will affect the resultant coating, and have a strong effect on cell capacity, rate capability, and aging behavior. Anode coating defects can lead to cell failure and cell thermal runaway.

Positive Electrode (Cathode)

There are varieties of positive electrode materials used in traditional lithium-ion cells—as with the negative electrode, these materials are powders that are combined with conductivity enhancers (carbon) and binder, and coated in a thin layer onto a current collector.[13] The most common cathode material in lithium-ion cells is lithium cobalt dioxide: a layered oxide material commonly referred to as "cobalt oxide."[14] However, various other materials are used such as lithium iron phosphate ($LiFePO_4$), spinels such as lithium manganese oxide ($LiMn_2O_4$), or mixed metal oxides that include cobalt (Co), nickel (Ni), aluminum (Al), and manganese oxides such as nickel cobalt aluminate (NCA) material ($LiNi_{0.8}Co_{0.15}Al_{0.05}O_2$) and nickel manganese cobaltite (NMC) material ($LiNi_{1/3}Mn_{1/3}Co_{1/3}O_2$). As with negative electrode materials, positive electrode materials can also vary dramatically based on source, purity, particle characteristics, coatings on particles, use of dopants, mixture ratios of various components, degree of compaction, crystallinity, etc.

A number of studies have attempted to rate the "safety" of different positive electrode materials.[15,16] These studies are based on thermal stability measurements of the cathode materials with electrolyte at full-charge voltage conditions. These tests show that cathode materials begin to react exothermically with electrolyte at a range of temperatures from approximately 130–250°C (270–480°F). Safety rankings based on this data have been strongly criticized in the industry because they relate to only a single aspect of cell safety: the reactivity of the cathode. They do not take into account the many other factors that contribute to cell safety such as the reactivity of the anode (which usually begins to react exothermically at much lower temperatures), cell construction details that may affect the likelihood

---

[13] For a detailed discussion of oxide cathode materials, see: Goodenough JB, "Oxide Cathodes," *Advances in Lithium-Ion Batteries*, WA van Schalkwijk and B Scrosati (eds), Kluwer Academic/Plenum Publishers, NY, 2002.

[14] Pillot C, "Present and Future Market Situation For Batteries," Proceedings, Batteries 2009, September 30–October 2, 2009, French Riviera; Pillot C, "Main Trends for Rechargeable Battery Market 2009–2020," Proceedings, Batteries 2010, September 29–October 1, 2010, French Riviera.

[15] Jiang J, Dahn J, *Electrochem. Comm.* **6**, 1, 39–43, 2003.

[16] Takahashi M, Tobishima S, Takei K, Sakurai Y, *Solid State Ionics*, **3–4**, 283–298, 2002.

of developing an internal short within the cell, the probability of manufacturing defects to cause internal shorting, etc.

Electrolyte

The electrolyte in a lithium-ion cell is typically a mixture of organic carbonates such as ethylene carbonate or diethyl carbonate (see Table 1.1 for flammability and auto-ignition temperatures of common carbonates used in lithium-ion cell electrolytes). The mixture ratios vary depending upon desired cell properties (e.g., a cell designed for low-temperature applications will likely contain a lower viscosity electrolyte than one optimized for room temperature applications). These solvents contain solvated lithium-ions, which are provided by lithium salts, most commonly lithium hexafluorophosphate ($LiPF_6$). Cell manufacturers typically include low concentrations of a variety of additives to improve performance characteristics such as overcharge resistance, cycle life, calendar life, and cell stability.[17] Gelling agents are added to the electrolytes of some pouch cells to mitigate the results of pouch puncture[18] and, in some instances, physically bind the electrodes together.

At typical cell voltages, mixtures of lithiated carbon (or lithium metal) and organic electrolyte are not thermodynamically stable and a reaction between the two materials will occur. Near room temperature conditions, the result of this reaction is the formation of a passivating layer on the carbon surface, commonly referred to as the solid electrolyte interphase (SEI) and some gases that result from breakdown of the electrolyte (short chain hydrocarbons, carbon dioxide, etc.).[19] During cell manufacturing, after cell assembly, the cell is slowly charged (and possibly repeatedly cycled and aged) during what is called "cell formation." This formation process is designed to produce a uniform and stable SEI layer on the cell anode. Note that formation is an exothermic process and the gases produced are usually flammable. The authors are unaware of publically available data on the specific flammability of gases produced during formation (these gases will be

---

[17] For a detailed discussion of electrolytes, see: Yamaki J-I, "Liquid Electrolytes," *Advances in Lithium-Ion Batteries*, WA van Schalkwijk and B Scrosati (eds), Kluwer Academic/Plenum Publishers, NY, 2002.

[18] For a detailed discussion of gelled electrolytes, see: Nishi Y, "Lithium-Ion Secondary Batteries with Gelled Polymer Electrolytes," *Advances in Lithium-Ion Batteries*, WA van Schalkwijk and B Scrosati (eds), Kluwer Academic/Plenum Publishers, NY, 2002.

[19] For a detailed discussion of the roll of SEI and other surface films, see: Aurbach D, "The Role of Surface Films on Electrodes in Li-Ion Batteries," *Advances in Lithium-Ion Batteries*, WA van Schalkwijk and B Scrosati (eds), Kluwer Academic/Plenum Publishers, NY, 2002.

Table 1.1 Measured flash points, auto-ignition temperatures, and heats of combustion of some typical lithium-ion cell organic electrolyte components

Electrolyte component	CAS registry number	Molecular formula	Melting point[a]	Boiling point[a]	Vapor pressure (torr)[b]	Flash point[b]	Auto-Ignition temperature[b]	Heat of combustion[c]
Propylene carbonate (PC)	108-32-7	$C_4H_6O_3$	−49°C −56°F	242°C 468°F	0.13 at 20°C	135°C 275°F	455°C 851°F	−20.1 kJ/ml −4.8 kcal/ml
Ethylene carbonate (EC)	96-49-1	$C_3H_4O_3$	36°C 98°F	248°C 478°F	0.02 at 36°C	145°C 293°F	465°C 869°F	−17.2 kJ/ml −4.1 kcal/ml
Di-Methyl carbonate (DMC)	616-38-6	$C_3H_6O_3$	2°C 36°F	91°C 195°F	18 at 21°C	18°C 64°F	458°C 856°F	−15.9 kJ/ml −3.8 kcal/ml
Diethyl carbonate (DEC)	105-58-8	$C_5H_{10}O_3$	−43°C 45°F	126°C 259°F	10 at 24°C	25°C 77°F	445°C 833°F	−20.9 kJ/ml −5.0 kcal/ml
Ethyl methyl carbonate (EMC)	623-53-0	$C_4H_8O_3$	−14°C 6.8°F	107°C 225°F	27 at 25°C	25°C 77°F	440°C 824°F	None available

[a] *CRC Handbook of Chemistry and Physics*, 91st Edition, Internet version 2011, Haynes WM (ed-in-chief), Lide DR (ed), Chapter 3

[b] Values of vapor pressure, flash point (closed cup), and auto-ignition temperatures are from MSDS of different sources. Note that the values are slightly different from different sources

[c] Harris SJ, Timmons A, Pitz WJ, "A Combustion Chemistry Analysis of Carbonate Solvents Used in Li-ion Batteries," *Journal of Power Sources*, 193 (2009), pp. 855–858

composed of decomposition products of original electrolyte solvents). Limited data is available in the literature regarding the composition of gases produced during formation. For example, in experiments concerning gas generation during formation of lithium-ion cells, Jehoulet et al.,[20] of SAFT detected the formation of ethylene and propylene gas, as well as small quantities of hydrogen, oxygen, nitrogen, carbon monoxide, methane, and carbon dioxide from cells that incorporated an electrolyte composed of propylene carbonate (PC), ethylene carbonate (EC), and dimethyl carbonate (DMC). Sandia National Laboratories (Sandia) has conducted gas analysis from punctured cells not subject to thermal runaway reactions.[21] Tested cells were produced by Quallion and had nickel cobalt aluminate cathodes (NCA material), and an electrolyte composed of $LiPF_6$ in a mixture of ethylene carbonate (EC) and ethyl-methyl carbonate (EMC). Sandia tested a fresh cell at 100% state-of-charge (SOC) and a cell that had been aged at 80% SOC at 45°C (113°F) for 8 weeks (gas was sampled from this cell at 100% SOC). Results of this testing are shown in Table 1.2. The observed argon, nitrogen, and oxygen likely remained from the cell assembly process. Electrolyte solvent (EC/EMC mixture) was detected in significant quantity. Electrolyte decomposition products from the cell formation and aging processes (H_2, CO, CO_2, methane, and ethylene) were also observed.

As temperature increases, reaction rates between the electrolyte and lithiated carbon increase exponentially (following Arrhenius behavior). Thus, lithium-ion cell capacity fades and internal impedance growth accelerates with increased ambient temperatures; most lithium-ion cells are not designed to be operated or stored above approximately 60°C (140°F). Many soft-pouch cell designs exhibit swelling if operated or stored at 60°C or above, due to gas generation from reactions similar to those responsible for SEI-formation.

For most commercial lithium-ion chemistries, the SEI layer itself will breakdown when cell temperature reaches the range of 75–90°C (167–194°F; exact temperature depends upon cell chemistry and SOC). Accelerated rate calorimetery (ARC) has shown that commercial lithium-ion cells exhibit self-heating behavior if brought to a temperature of about 80°C (176°F).[22] If cells are then maintained in an adiabatic environment (e.g., if they are well insulated), the cells can then self-heat to thermal runaway conditions (this process requires approximately 2 days for an 18650 cell tested in an ARC). Note that United Nations (UN) and Underwriters

---

[20] Jehoulet C, Biensan P, Bodet JM, Broussely M, Moteau C, Tessier-Lescourret C, "Influence of the solvent composition on the passivation mechanism of the carbon electrode in lithium-ion prismatic cells," Proceedings, Symposium on Batteries for Portable Applications and Electric Vehicles, 1997.

[21] Roth EP, Crafts CC, Doughty DH, McBreen J, "Advanced Technology Development Program for Lithium-Ion Batteries: Thermal Abuse Performance of 18650 Li-Ion Cells," Sandia Report: SAND2004-0584, March 2004.

[22] White K, Horn Q, Singh S, Spray R, Budiansky N, "Thermal Stability of Lithium-ion Cells as Functions of Chemistry, Design and Energy," Proceedings, 28th International Battery Seminar and Exhibit, Ft. Lauderdale, FL, March 14–17, 2011.

Table 1.2 Gas composition of punctured cells from Sandia testing

Cell type	Fresh cell at 100% SOC	Aged cell[a] at 100% SOC
Max sample temperature	25°C	45°C
Gas species	Volume percent (%)	
H_2	8.2	0.3
Argon	44.0	27.8
N_2	6.2	9.6
O_2	0.1	1.7
CO	4.2	11.3
CO_2	12.6	26.3
CH_4	13.5	11.5
C_2H_4	3.1	None detected
C_2H_6	None detected	None detected
Ethyl fluoride	None detected	None detected
Propylene	None detected	None detected
Propane	None detected	0.06
Electrolyte solvent (EC/EMC mixture)	11.2	11.5

[a] Cell was aged by being held at 45°C and 80% SOC for 8 weeks

Laboratories (UL) tests for lithium-ion batteries discussed below require cells exhibit long-term thermal stability in the range of 70–75°C (158–167°F).

The most commonly used electrolyte salt ($LiPF_6$) will decompose to form hydrofluoric acid (HF) if mixed with water or exposed to moisture. Cell production and assembly is conducted in "dry rooms" to prevent HF formation (the presence of HF in cells will cause degradation of the cells). Leakage of free electrolyte from cells can result in deposition of the electrolyte salt as organic components volatilize.

Electrolyte chemistry is an active area of research. A number of groups have conducted research to produce non-flammable, or reduced flammability electrolytes either through the addition of additives to typical organic solvent mixtures (see footnote 17), or through the development of non-organic ionic liquids.[23] Researchers have also attempted to produce electrolytes suited to low temperature applications,[24] and have experimented with salts other than $LiPF_6$ (see footnote 1). However, at the time of this writing, none of these electrolytes have proven to be widely commercially viable and are not common in the field.

[23] Webber A, Blomgren GE, "Ionic Liquids for Lithium-Ion and Related Batteries," *Advances in Lithium-Ion Batteries*, WA van Schalkwijk and B Scrosati (eds), Kluwer Academic/Plenum Publishers, NY, 2002.

[24] Smart MC, Ratnakumar DV, Chin KB, Whitcanak LD, Lithium-Ion Electrolytes Containing Ester Cosolvents for Improved Low Temperature Performance, J. Electrochem. Soc., **157**(12), (2010), pp. A1361–A1374 (2010).

Fig. 1.11 An example of a micro-shorting location on a separator, at the point of shorting, the separator locally melted and shutdown. The micro-short is approximately 1 mm in diameter

Separator

Lithium-ion cell separators most commonly are porous polyethylene, polypropylene, or composite polyethylene/polypropylene films.[25] These films are typically on the order of 20 μm thick, although thinner (approximately 10 μm) and thicker films can be found (approximately 40 μm). The function of the separator is to prevent direct contact between the anode and cathode. The pores in the separator allow transfer of lithium ions by diffusion during charge and discharge. These films soften and close their pores at elevated temperatures (usually in the range of 130–150°C/270–300°F), and stop charge or discharge processes by impeding the transport of ions between the anode and cathode. Thus, these types of separators are commonly referred to as "shutdown" separators. If a minor internal short occurs within a cell (e.g., from small contaminants penetrating the separator), local separator shutdown will effectively disable a small point within the cell by melting slightly and closing the separator pores (Fig. 1.11). The shutdown function will also permanently disable the entire cell in the case of an abnormal internal temperature rise to approximately 130°C (266°F) (e.g., due to high current draws caused by an external short circuit of the cell) (Fig. 1.12). However, should internal temperatures rise significantly above approximately 150°C (300°F) the separator will melt entirely and allow contact between the anode and cathode. Figure 1.13 uses differential scanning calorimetry (DSC) to graphically illustrate the thermal transitions of a typical separator material.

Separator thickness, porosity, permeability, toughness, and resistance to penetration can vary considerably depending on desired cell properties. For example, one way to increase the capacity and rate capability of a cell design, is to select a thinner separator thus including more electrode material in a given, fixed, cell case.

---

[25] http://www.celgard.com/products/default.asp

Fig. 1.12 An example of separator melting due to electrical abuse of a cell

However, it is generally known that this strategy can also lead to cell failures, as thinner separators can be more susceptible to damage. In the past, some cell manufacturers found cell failure rates increased significantly when the separator was made thinner. Separator characteristics are measured using a number of ASTM standard test methods developed for characterizing plastic sheets and films, as well as industry specific methods developed by research laboratories[26] and separator manufacturers.[27] UL has developed a standard approach to characterizing separator material described in UL Subject 2591, "Battery Separators" that specifies methods for characterizing separator construction and performance properties such as permeability, tensile strength, puncture strength, dimensional stability, shutdown temperature, and melting temperature.

New separators continue to be developed and applied to commercial cells. Some separator manufacturers are currently producing or experimenting with separators that incorporate ceramic coatings or separators made of thermally stable non-woven fabrics that do not have shutdown capability but maintain separation between the anode and cathode over a broader temperature range.[28]

---

[26] Zhang SS, "A review on the separators of liquid electrolyte Li-ion batteries," *Journal of Power Sources*, **164** (2007), pp. 351–364.

[27] Arora P, Zhang Z, "Battery Separators," *Chemical Reviews*, **104** (2004), pp. 4419–4462.

[28] Roth EP, Doughty DH, Pile DL, "Effects of separator breakdown on abuse response of 18650 Li-ion cells," Journal of Power Sources, **174** (2007), pp. 579–583.

Fig. 1.13 Differential scanning calorimetry (DSC) showing melting endotherms at 133 and 159°C for a typical polyethylene/polypropylene separator material

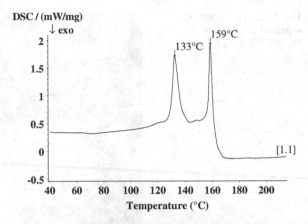

Fig. 1.14 Current collector foils prior to coating with active material

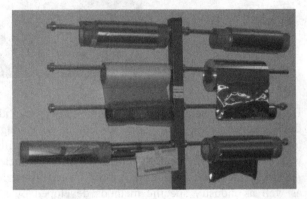

Current Collectors

The most common current collectors are thin foils of copper (used as a substrate for anode active materials) and aluminum (used as a substrate for cathode active materials) (Figs. 1.14 and 1.15). The role of the current collector is to transfer current evenly throughout the cell to the active material, to provide mechanical support for the active material, and to provide a point of mechanical connection to leads that transfer current into the cell (internal leads may be welded to regions of bare current collector).

The use of copper as the current collector for the negative electrode has particular reliability and safety implications. At very low cell voltages (usually approximately 1 V for the cell), the potential at the copper current collector increases to the point where copper will begin to oxidize and dissolve as copper ions into the electrolyte. On subsequent recharge, the dissolved copper ions plate as copper metal onto negative electrode surfaces, reducing their permeability and making the cell susceptible to lithium plating and capacity loss. Usually, once a severe over-discharge event has occurred, cell degradation accelerates: once the

Fig. 1.15 Layers of material from a wound cylindrical cell; left to right: negative electrode (graphite coated onto copper), separator, positive electrode (metal oxide coated onto aluminum), and separator

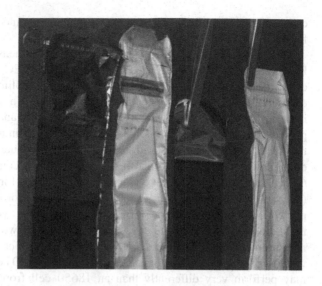

negative electrode has become damaged by copper plating it will no longer be able to uptake lithium under "normal" charge rates. In such an instance, "normal" charge cycles cause lithium plating, which result in a greater loss of permeability of the surfaces. Ultimately, over-discharge of cells can lead to cell thermal runaway.

Most consumer electronics devices set specific discharge limits for their lithium-ion battery packs to prevent over-discharge. The protection electronics disconnect the pack from the discharge load once any individual series element voltage drops below a specified cut-off. This protection is effective for normally operating cells but it cannot prevent over-discharge resulting from an internal cell fault and self-discharge of the cells. Thus, if a device is fully discharged and then stored for an extended period, the cells may become over-discharged, or if a mild short exists within the battery, the cells may become over-discharged within a short time. Most battery pack protection electronics allow recharge of over-discharged cells, despite the potential for the negative electrode becoming damaged. In single cell consumer applications (e.g., cell phones), the resulting capacity fade, and elevated impedance of the battery generally drives a user to replace the battery pack. Nonetheless, over-discharge does periodically cause thermal runaway of single cell battery packs. In multi-series element battery packs (e.g., notebook computers), capacity fade and elevated impedance usually causes a severe block imbalance that drives permanent disabling of the battery pack.

Cell Enclosures (Cases and Pouches)

Cells can be constructed in a variety of form factors and materials. Generally, cell form factors are classified as cylindrical, prismatic (flat rectangle), and pouch cells (also known as lithium-ion polymer, soft-pack polymer, lithium polymer, or Li-Po cells). Figure 1 of Preface shows some typical commercial electronic cell form factors.

Cells are most often designated based on their dimensions per an International Electrotechnical Commission (IEC) Standard CEI/IEC 61960.[29] For cylindrical cells, the first two digits define cell diameter in millimeters and the next three digits define cell length in tenths of millimeters. Thus, the 18650 designation indicates a cylindrical cell with a diameter of 18 mm, and a length of 65.0 mm. At present, the 18650-size cell is the most common cylindrical cell size. Cells with the 18650-form factor are used in most laptop computer batteries and numerous other devices. The Tesla Roadster battery pack is composed of approximately 6,800 18650-cells.[30] Another common cylindrical cell form factor is the 26650 cell (26 mm diameter, 65.0 mm length). Cells with this form factor are often used in power tool applications. For prismatic cells, the first two digits define cell thickness, the next two designate cell width, and the last two designate cell length: all measurements are in millimeters. Note that a form factor-based designation does not describe cell chemistry or capacity. Thus, an 18650 cell from one manufacturer may perform very differently than an 18650 cell from a second manufacturer. Manufacturers may include a variety of other codes with cell size designations to describe their products. At present, these other codes are manufacturer specific and may not follow a standardized designation.

Hard case cells have an enclosure composed of metal: usually nickel-coated steel or aluminum. Generally, the enclosure of a hard case cell is one of the cell electrodes, and leads can be directly spot welded to the case: for nickel-coated steel cases (18650 cells) the case is negative; for aluminum cases (many prismatic cells) the case is positive. Since these cases are polarized, they are usually at least partially covered with shrink-wrap to provide electrical isolation. To minimize the likelihood of cell leakage, designers attempt to minimize the number of case seams. Thus, these cases are usually "deep drawn" cans that only require formation of a seal at one end cap. The end cap closure is accomplished either with gaskets (typical of 18650 cells) (Fig. 1.16) or with welds (Fig. 1.17). In order to allow for safe venting[31] should a cell become over-pressurized, hard case designs require the inclusion of a safety vent. Vents are usually formed by including a burst disk in the cell design (typical in an 18650 design) (Fig. 1.18), by including a score mark on the cell (typical in prismatic designs), or by adjusting weld strength to allow failure of weld closures at safe venting pressures. Since hard cases provide mechanical protection to cell electrodes, they can be relatively densely packed for shipping purposes. In addition, dense packing arrangements can be

---

[29] CEI/IEC 61960 2003-12, Secondary cells and batteries containing alkaline or other non-acid electrolytes—Secondary lithium cells and batteries for portable applications, International Electrotechnical Commission.

[30] http://webarchive.teslamotors.com/display_data/TeslaRoadsterBatterySystem.pdf

[31] Should a safety vent not operate properly, on thermal runaway a cell case could rupture at an elevated pressure and distribute cell materials over a wide radius, such rupture is sometimes called "rapid disassembly."

Fig. 1.16 Cap assembly cross section of an 18650 cell with sealing gasket indicated

Fig. 1.17 Laser welding is commonly used to seal hard case prismatic cells

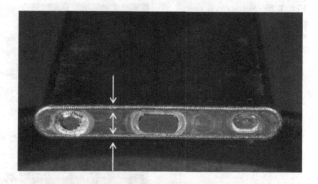

Fig. 1.18 Cap assembly cross section of an 18650 cell with burst disk indicated

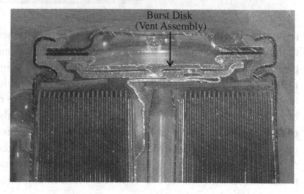

used in battery pack designs, and packs may not need to supply additional mechanical protection; thus, it is not uncommon to encounter small 2- or 4-cell packs that consist of cells merely shrink-wrapped together, and electrically connected to protection electronics (Fig. 1.19).

Soft-pouch cells (also commonly referred to as pouch, polymer, or Li-Po cells) have an enclosure of polymer-coated aluminum foil. This type of enclosure allows production of light and very slender cell designs that are not possible to make in a hard case format. Seams of the enclosure are heat-sealed. Vent ports do not need to be included in soft-pouch cells, as the seams will fail at relatively low pressures and

Fig. 1.19 Examples of 2-
and 4-cell packs composed of
18650 cells shrink-wrapped
together

Fig. 1.20 Soft-pouch cells
placed in molded tray and
ready for transport

temperatures. Pouches are designed to be electrically neutral. Thus, all connections
to the cell must be made at leads protruding from the pouch. Should a pouch become
polarized it will likely corrode and result in cell leakage and swelling: a common
failure mode for soft-pouch cells. Since pouches provide limited mechanical protection to cell electrodes, mechanical protection of the cells must be accomplished
by surrounding materials. When bare soft-pouch cells are transported, they are
placed in molded trays that separate individual cells (Fig. 1.20). When included in a
product, a soft-pouch cell may be embedded in a device and the device case itself
may provide mechanical protection. Alternatively, a pouch cell may be enclosed in a
metal sleeve with plastic end caps (common in cell phone applications).

Charge Interrupt Devices

Because overcharge leads to thermal runaway in lithium-ion cells, many cell
designs include built-in mechanisms to prevent overcharge. Overcharge can lead
to significant gas generation within a cell prior to the cell entering a thermal
runaway condition (see footnote 19 in Chap. 4). In prismatic form factors, and
particularly in cells with thin cases or with soft-pouch cells, gas generation within

Fig. 1.21 Cap assembly cross section of an 18650 cell with CID assembly weld point indicated (*circle*)

the cell will result in cell swelling and may force electrodes apart, effectively curtailing the transfer of ions and interrupting charging. This process can prevent thermal runaway of the cells, but is not always effective.[32]

The geometry of cylindrical cells prevents separation of electrodes if gas generation occurs. Cell designers have developed mechanical charge interrupt devices (CIDs) for cylindrical cells used in consumer electronic devices (Fig. 1.21). On activation, CIDs physically and irreversibly disconnect the cell from the circuit. Although CIDs are usually described as overcharge protection devices, they will activate if anything causes cell internal pressure to exceed the activation limit. This could include overcharge, cell overheating, significant lithium plating followed by electrolyte breakdown, mild internal shorting, and/or significant cell over-discharge. Proper design and installation is required for reliable operation of CIDs. CIDs must also be appropriately matched to cell chemistry so that overcharge conditions result in sufficient gas generation prior to thermal runaway to activate the CID. If a CID is not properly matched to cell chemistry, low current overcharge or very high over currents may not activate a CID sufficiently early to prevent cell thermal runaway.

Due to their design, traditional CIDs may not be applicable to very high rate cells such as those used in power tools, because the traditional CID design will not allow transfer of very high currents. In addition, CIDs may not be appropriate for application to large parallel arrays of cells. In 2- or 3-cell parallel arrays, CIDs generally work as expected and facilitate a graceful failure of a battery pack. However, it is unlikely that all CIDs in a large parallel array of cells will activate simultaneously, but rather, CID activation will occur in a cascade causing high over currents to be applied to cells where CIDs have not activated. Rapid application of high currents may drive cells to thermal runaway before their CIDs can activate.[33]

---

[32] Exponent observed despite pouch swelling behavior it remains possible to drive prismatic and pouch cells into thermal runaway.

[33] Jeevarajan J, "Performance and Safety Tests on Lithium-Ion Cells Arranged in a Matrix Design Configuration," Space Power Workshop, The 2010 Space Power Workshop, Manhattan Beach, CA, April 20–22, 2010.

Fig. 1.22 Cap assembly
cross section of an 18650 cell
with PTC device indicated

Positive Temperature Coefficient Switches

High rate discharges can cause heating of cells, in some cases to the point of damaging internal components such as the separator, and can lead to cell thermal runaway. Polymeric positive temperature coefficient (PTC) devices, also called resettable thermistor devices, or "polyswitches" are common components of commercial cells (e.g., part of the cap assembly of 18650 commercial cells) (Fig. 1.22) or commercial battery packs (placed in the circuits of battery packs designed with prismatic cells). These devices include a conductive polymer layer that becomes very resistive above some threshold temperature. PTC devices are selected to remain conductive within specified current and temperature conditions. However, should discharge (or charge) current become excessive, the polymer will heat and become highly resistive, greatly reducing current from (to) the cell. Once the PTC device cools, it again becomes conductive. PTC devices may not be applicable to high current cells (e.g., power tool cells) or battery packs composed of high numbers of cells connected in parallel (see footnote 33).[34]

Battery Pack Protection Electronics

Lithium-ion batteries require relatively complex protection circuitry to protect against a variety of electrical abuse scenarios including over voltage overcharge, over current overcharge, discharging at an excessive current (external short circuit), charging outside an acceptable temperature range, over discharge, and

---

[34] Smith K, Kim GH, Darcy E, Pesaran A, "Thermal/electrical modeling for abuse-tolerant design of lithium ion modules," International Journal of Energy Research, 34 (2010), pp. 204–215.

imbalance protection for multi-series battery packs.[35,36] UN and various commercial standards such as UL, IEEE, and automotive standards (discussed in Chap. 3) govern the minimum functionality requirements for protection electronics systems.

The protection circuitry for consumer electronics applications is typically integrated into a single PCB that achieves most or all of the protection functions discussed above. This PCB is sometimes referred to as the pack protection PCB or as the battery management unit (BMU). Some device designs move some or all protection functionality to the host device rather than using a dedicated PCB in the battery. Protection electronics for electric vehicles can be divided among modules, with coordination between modules occurring at the pack level. In the electric vehicle community, the full system of pack protection electronics is usually referred to as the battery management system or BMS.

Maximum charge voltage for a lithium-ion cell varies depending on the specific battery chemistry or the intended use environment (4.2 V is typical for many chemistries). Preventing overcharge is considered sufficiently critical to warrant individual monitoring of cell or series element (block) voltages by electronics to prevent any cell from exceeding a voltage limit. In addition, most protection electronic packages include multiple independent circuits to terminate charge so that a single-point circuitry failure cannot disable over-voltage protection.

Charging at too high of a current can lead to conditions resembling overcharge of lithium-ion cells (over current overcharge), or can cause heating at connections both internal and external to cells leading to undesirable effects (see footnote 36). Most lithium-ion batteries contain electronics to regulate charging currents. Normal charge rates are usually set to be some fraction of their maximum allowed cell charging rate: a safety margin is typically provided to account for aging of cells. Charging lithium-ion batteries at low temperatures can result in lithium plating due to reduced lithium ion diffusion rates within the negative electrode. Charging or discharging lithium-ion batteries at high temperature can increase the risk of significant gas generation within the batteries, leading to swelling, the nuisance operation of pressure-triggered protective devices (e.g., CIDs) or thermal runaway due to mechanical disturbance of windings or layers. Thus, lithium-ion battery packs often include controls to prevent charging at excessively low or high temperatures.

Over-discharging lithium-ion cells can cause damage to current collectors, and ultimately electrodes, leading to compromised performance or increased risk of thermal runaway. Thus, protection circuits tend to prevent over-discharge and can be designed to go into low power sleep modes below certain specified voltages. In

---

[35] For a detailed discussion of charging algorithms see: van Schlakwijk WA, "Charging, Monitoring and Control," *Advances in Lithium-Ion Batteries*, WA van Schalkwijk and B Scrosati (eds), Kluwer Academic/Plenum Publishers, NY, 2002.

[36] For a more detailed discussion of lithium-ion protection electronics design see: Friel DD, "Battery Design," *Linden's Handbook of Batteries*, 4th Edition, TB Reddy (ed), McGraw Hill, NY, 2011.

some instances, a low voltage condition, above the threshold for current collector damage, is tolerable. A lithium-ion cell in such a state of deep discharge will likely require low charging currents until the cell reaches some threshold voltage. Thus, lithium-ion battery packs often include controls to limit charge currents until a desired voltage threshold is reached.

In multi-series element battery packs, cells in the various series elements may not age uniformly, resulting in divergent capacities among series elements. Individual series element voltage sensing is used to prevent over charge or over-discharge of any element. However, a significant imbalance can indicate a problematic cell, or lead to over current damage of a high impedance series element. Thus, protection circuits for multi-series battery packs (especially for notebook computers) often include the capability to permanently disable a battery pack in which imbalance has become too severe.

Battery Pack Enclosures

Battery pack enclosures can vary considerably, and will depend upon the application. Some of the simplest pack enclosures (for hard case cells) are simply a layer of shrink-wrap that holds cells together. Notebook computer battery pack enclosures can consist of hard plastic cases or hard metal cases. If hard case cells are used, a notebook battery pack enclosure may consist of a heavy plastic case on five sides with a heavy decal over a plastic frame on the sixth side that is usually enclosed by the notebook computer itself. Enclosures for pouch cells used in consumer electronics applications generally include stiff metal and plastic members (or metal sheets embedded in plastic members) to protect the pouch cells from mechanical damage. However, pouch cells used for remote control model aircraft may be simply wrapped in shrink-wrap. Large format module and battery pack enclosures are currently being developed. Note that if plastic is used for the battery enclosure, it could contribute significantly to heat release if a fire were to occur.

Chapter 2
Lithium-Ion Technology Applications

Lithium-ion cells have gained a dominant position in the rechargeable battery market for consumer electronic devices.[1] Market research data (see footnote 14 in Chap. 1) indicates the lithium-ion cell market is growing by approximately 20% per year, while the nickel metal hydride (NiMH) battery market has stagnated (or only grown slightly due to increased demand for HEV vehicles), and the nickel cadmium (NiCad) market has a negative annual growth rate of 16%. Lithium-ion technologies have almost entirely displaced other chemistries in cell phone and notebook computer applications. Lithium-ion cells have begun to displace NiCad and NiMH cells in power tools and household products such as remote controls, mobile phones, cameras, and some toys.

The primary reason for lithium-ion battery dominance is the chemistry's high specific energy (Wh/kg) and high energy density (Wh/L), or more simply, the fact a lithium-ion cell of a specific size and weight will provide substantially more energy than competing technologies of the same size or weight. Lithium-ion cells have enabled smaller, more slender, and more feature rich portable electronics designs. Now that lithium-ion cells are readily available and cost has decreased, designers are more likely to select this technology for a wide range of applications. For example, in 2010, Best Buy Corporation[2] estimated they had approximately "12,000 active SKU's of consumer electronics and appliances" many of which contained lithium or lithium-ion batteries. Best Buy estimated that products containing lithium-ion batteries included: portable GPS devices, portable game players, portable DVD players, portable TVs, portable radios, cell phones, music players, e-readers, notebook computers, cordless headphones, universal remote

---

[1] For a more detailed discussion of lithium-ion cells in consumer electronic devices see: Wozniak JA, "Battery Selection for Consumer Electronics," *Linden's Handbook of Batteries*, 4th Edition, TB Reddy (ed), McGraw Hill, NY, 2011.
[2] PHMSA-2009-0095-0112.1.

C. Mikolajczak et al., *Lithium-Ion Batteries Hazard and Use Assessment*,
SpringerBriefs in Fire, DOI: 10.1007/978-1-4614-3486-3_2,
© Fire Protection Research Foundation 2011

controls, cameras, camcorders, two-way radios, rechargeable vacuums, electric razors, electric toothbrushes, electric vehicles, and more.

Many small devices implement only a single lithium-ion cell (3–4 V systems) with fairly rudimentary protection electronics. The smallest lithium-ion cells are found in devices such as hearing aids,[3] Bluetooth headsets,[4] and very small MP3 players.[5] Very small cells are also being implemented in medical devices such as part of sensor packages that can be attached to the human body and allow patient monitoring.[6] Some highly specialized implantable lithium-ion batteries are also available (see footnote 9 in Chap. 1).[7] Larger single cell applications include batteries for digital cameras, MP3 players, and e-readers. The most common single cell lithium-ion battery application is cell phones and smart-phones. As a result, for most single cell applications, designers follow recommendations set forth in IEEE 1725, "Standard for Rechargeable Batteries for Cellular Telephones," and apply battery protection electronics hardware developed for cell phone applications.

For larger electronic devices such as notebook computers, power tools, portable DVD players, and portable test instruments,[8] multi-cell battery packs are used. Multi-cell devices such as notebook computer battery packs run at nominal voltages of 14.4 V with capacities up to 6.6 Ah (95 Wh), and utilize complex protection electronics. Notebook computers represent the largest population of relatively complex lithium-ion batteries in the commercial market. Most of these packs contain between six and twelve 18650 cells connected in series and parallel: the most common pack configuration involves 3- or 4-series elements, each element consisting of blocks of two cells connected in parallel (3s, 2p or 4s, 2p packs). As a result, for most consumer electronics, multi-cell applications, designers follow recommendations set forth in IEEE 1625, "Standard for Rechargeable Batteries for Multi-Cell Mobile Computing Devices," and apply battery protection electronics hardware developed for notebook computer applications.

There are some notebook computer battery packs and power tool battery packs that include larger cells or higher cell counts. However, the size of commercially available battery packs has been effectively limited by international shipping regulations,[9] which provide exemptions to hazardous materials transport rules for lithium-ion cells smaller than 20 Wh (effectively a 5-Ah cell with a nominal

---

[3] http://www.cochlear.com/au/nucleus-cochlear-implants/nucleus5/battery-choices

[4] http://www.jawbone.com/headsets/era/specs

[5] http://store.apple.com/us/browse/home/shop_ipod/family/ipod_shuffle?afid=p219%7CGOUS& cid=AOS-US-KWG

[6] http://www.mindray.com/en/products/9.html

[7] For a more detailed discussion of lithium-ion cells in medical applications see: Leising RA, Gleason NR, Muffoletto BC, Holmes CF, "Batteries for Biomedical Applications," *Linden's Handbook of Batteries*, 4th Edition, TB Reddy (ed), McGraw Hill, NY, 2011.

[8] http://www.fluke.com/fluke/usen/portable-oscilloscopes/fluke-190-series-ii-scopemeter.htm?PID= 70366

[9] UN Transport of Dangerous Goods—Model Regulations, ICAO Technical Instruction for the Safe Transport of Dangerous Goods by Air, IATA Dangerous Goods Regulations, etc.

voltage of 3.7 V) and lithium-ion batteries smaller than 100 Wh (e.g., a battery pack with twelve 18650 cells of 2.2-Ah capacity each). Cells or battery packs that fall outside of the exemption limits must be transported as hazardous materials.

Some relatively small number of larger (large format[10]) lithium-ion battery packs have been manufactured for certain low volume (at the time of this writing) applications such as Segway personal transporters,[11] electric bicycles,[12] electric scooters, electric vehicles, commercial aircraft auxiliary power units (APUs), satellites, military applications, and energy storage and grid stabilization applications. Some of these large battery packs have been constructed using cells common to commercial applications (e.g., the Tesla Roadster battery pack is constructed from approximately 6,800 18650 cells[13]). These designs involve connecting ten or more cells in parallel to form elements or blocks that are then connected in series. Other large battery packs have been constructed from "large format cells" that have capacities in the range of 10–100 Ah. Standards for these sorts of applications are currently being written or revised to be appropriate for lithium-ion technology (discussed below). In addition, pack protection electronics hardware for these high voltage and high current applications remains in the development phase, with limited hardware that can be bought "off the shelf."

The demand for hybrid electric vehicles (HEVs), plug-in hybrid electric vehicles (PHEVs), and purely electric vehicles (EVs) is expected to increase. At present, many hybrid vehicles (e.g., Toyota Prius, Ford Escape) implement NiMH batteries. A few vehicles that implement lithium-ion battery technology have recently entered the U.S. market including the Tesla Roadster, Nissan Leaf, and Chevrolet Volt. Lithium-ion battery powered vehicles have also entered overseas markets. For example, in China, lithium-ion batteries have been adapted for use in buses and automobiles such as taxicabs.[14] Lithium-ion technology is expected to penetrate this market and achieve approximately 35% market share by 2020; NiMH batteries are expected to continue to dominate the market through 2020.[15] Current projections suggest that lithium-ion batteries will come to dominate the PHEV and EV markets, while NiMH batteries will remain dominant in HEV

---

[10] The term "large format" is loosely applied in the Li-ion battery area, as the definition is linked to transport regulatory requirements that have been subject to change. Based on recent UN Model Regulations, a large format cell contains more than 20 h of energy (e.g., more than 5 Ah capacity with a 3.7 nominal voltage), while a large format battery pack contains more than 100 Wh of energy (e.g., a battery pack containing more than twelve 2.2 h cells).

[11] http://www.segway.com/individual/models/accessories.php#batteries

[12] http://www.pingbattery.com/servlet/StoreFront

[13] Staubel JB, "Safety Testing of Tesla's Battery Packs," Proceedings, Nineth International Advanced Automotive Battery & EC Capacitor Conference (AABC), June 2009.

[14] http://chinaautoweb.com/2011/04/hangzhou-halts-all-electric-taxis-as-a-zotye-langyue-multipla-ev-catches-fire/ and http://green.autoblog.com/2011/06/16/zotye-electric-taxi-fire-caused-by-shoddy-chinese-built-battery/

[15] Pillot C, "The Battery Market for HEV, P-HEV & EV 2010-2020," Proceedings, 28th International Battery Seminar & Exhibit, March 14-17, 2011, Ft. Lauderdale, FL.

markets. US market penetration of HEVs is expected to reach about 10% by 2015. Significantly lower market penetration rates are expected for EV and PHEV vehicles.[16] The FPRF has compiled a detailed report regarding battery placement and fire fighter safety for EV and HEV vehicles.[17]

With penetration of electric vehicles, comes addition of charging stations in public areas as well as in private residences. Automotive battery packs will also be serviced and thus, stored at service locations, and also battery switching locations such as those being demonstrated by Better Place.[18] This type of new infrastructure will pose high voltage and fire safety challenges in addition to those associated with lithium ion batteries themselves discussed in this report.

Lithium-ion batteries have begun to replace other battery chemistries in aerospace applications. For example, in 2007, the Boeing Corporation requested a waiver from the US Federal Aviation Administration[19] to allow use of lithium-ion batteries for powering a number of systems on the 787 Dreamliner commercial aircraft design including: the main and APU, flight control electronics, the emergency lighting system, and as an independent power supply for the flight recorder. Lithium-ion batteries are already being used on a variety of military aircraft.[20] Lithium-ion batteries are being installed in a range of space applications including satellites, research probes, and manned mission power supplies.[21] Both large format cells[22] and large format battery packs composed of 18650 cells[23] are being used in these applications.

---

[16] For a more detailed discussion of batteries for electric vehicles see: Corrigan DA, Alvaro M, "Batteries for Electric and Hybrid Vehicles," *Linden's Handbook of Batteries*, 4th Edition, TB Reddy (ed), McGraw Hill, NY, 2011.

[17] Grant CC, "Fire Fighter Safety and Emergency Response for Electric Drive and Hybrid Electric Vehicles," Fire Protection Research Foundation, May 2010.

[18] http://www.betterplace.com/the-solution-switch-stations

[19] Federal Aviation Administration, 14 CFR Part 25 [Docket No. NM375; Notice No. 25-07-10-SC], Special Conditions: Boeing Model 787-8 Airplane; Lithium Ion Battery Installation
http://rgl.faa.gov/Regulatory_and_Guidance_Library/rgSC.nsf/0/80b9e22f91f3ae59862572cd 007014041!OpenDocument&ExpandSection=-4

[20] http://www.quallion.com/sub-mmil-apu.asp

[21] Spotnitz R, "Scale-Up of Lithium-Ion Cells and Batteries," *Advances in Lithium-Ion Batteries*, WA van Schalkwijk and B Scrosati (eds), Kluwer Academic/Plenum Publishers, NY, 2002.

[22] For examples, see: http://www.yardney.com/Lithion/lithion.html
http://www.saftbatteries.com/MarketSegments/Space/tabid/152/Default.aspx
Smart MC, Ratnakumar BV, Whitcanack LD, Puglia FJ, Santee S, Gitzendanner R, "Life Verification of Large Capacity Yardney Li-ion Cells and Batteries in Support of NASA Missions," International Journal of Energy Research, 2010 (34), pp. 116–132.

[23] http://www.abslspaceproducts.com/

Considerable interest has been generated in the last 2–3 years for applying lithium-ion batteries for a variety of energy storage and grid stabilization (stationary) applications.[24] Prototype systems have been installed.[25,26] Megawatt scale systems typically include thousands of cells housed in shipping container-sized structures that can be situated on power utility locations.[27] These systems usually include integrated fire suppression in their installations.[28] Smaller systems have also been planned and are being delivered for evaluation purposes, particularly for use with renewable energy sources.[29] There is also interest in distributed power storage, down to individual home level. Discussions regarding smart grid applications include using automotive battery packs connected to the grid for temporary energy storage, and as emergency power supplies when power is unavailable. There is also considerable discussion in the industry regarding repurposing used or refurbished automotive battery packs for stationary applications such as home level power storage once the packs are no longer suitable for use in vehicles.[30] It remains to be seen whether refurbishment of packs will be practical or economical,[31] as cells must generally be well matched to provide good performance in battery packs, and aged cells are particularly difficult to match effectively. In addition, for refurbished pack safety, the issue of determining when a cell should be retired will need to be resolved.

Although applications for large format lithium-ion battery packs remain fairly niche at the time of this writing, considerable momentum has developed for using lithium-ion cells to replace NiCad and lead acid batteries. As with consumer electronics, the lighter weight and smaller size of lithium-ion batteries appeals to designers concerned with energy efficiency in transportation applications. The current battery chemistry for energy storage/stationary applications such as datacenter-scale uninterruptable power supplies (UPS) is lead acid. In this type of application, the system designer must make a choice between shorter life, lower maintenance valve regulated lead acid (VRLA), or higher maintenance, longer life

---

[24] Kamath H, "Integrating Batteries with the Grid," Proceedings, 28th International Battery Seminar & Exhibit, March 14-17, 2011, Ft. Lauderdale, FL.

[25] http://www.renewableenergyfocususa.com/view/11958/a123-delivers-44-mw-smart-grid-stabilization-systems/

[26] Gengo T, et al., "Development of Grid-stabilization Power-storage System with Lithium-ion Secondary Battery," Mitsubishi Heavy Industries Technical Review, 46(2), June 2009.

[27] http://gnes2010.rmtech.org/_includes/presentations/chu.pdf

[28] http://www.saftbatteries.com/SAFT/UploadedFiles/PressOffice/2010/CP_31-10_eng.pdf

[29] For example, see: http://www.saftbatteries.com/SAFT/UploadedFiles/PressOffice/2011/CP_08-11_en.pdf.
http://www.ourmidland.com/news/01bb863d-25fe-502c-aaab-db4667e02162.html

[30] http://green.blogs.nytimes.com/2011/04/27/a-second-life-for-the-electric-car-battery/

[31] Neubauer J, Pesaran A, "PHEV/EV Li-Ion Battery Second Use Project," NREL/PR-540-48018, April 2010, at http://www.nrel.gov/docs/fy10osti/48018.pdf.

flooded lead acid (FLA) batteries.[32] Lithium-ion's low self-discharge rates and minimal maintenance requirements are therefore appealing to designers concerned with energy storage applications.[33]

[32] http://www.apcdistributors.com/white-papers/Power/WP-30%20Battery%20Technology%20 for%20Data%20Centers%20and%20Network%20Rooms%20-%20Lead-Acid%20Battery%20 Options.pdf

[33] For a more detailed discussion of electrical energy storage applications see: Akhil AA, Boyes JD, Butler PC, Doughty DH, "Batteries for Electrical Energy Storage Applications," *Linden's Handbook of Batteries*, 4th Edition, TB Reddy (ed), McGraw Hill, NY, 2011.

Chapter 3
Summary of Applicable Codes and Standards

Historically, lithium-ion battery development has been significantly impacted by codes and standards developed by several organizations: the hazardous materials transport regulations developed by the UN, the consumer electronics safety standards developed by UL and, more recently by the Institute of Electrical and Electronics Engineers (IEEE), and the IEC. These standards continue to define safety performance for lithium-ion cells. A number of additional standards have recently been adopted or developed: there are standards that apply to lithium-ion batteries in specific jurisdictions (e.g., in Japan, China, or Korea). Currently, the automotive industry is in the process of drafting new standards or revising existing standards for application to lithium-ion batteries.

Finally, recycling and product stewardship regulations targeted at used batteries are becoming more common in the United States. At present, in California, all battery types must be recycled, and may not be disposed of as solid waste. A new mandate requires battery manufacturers develop collection programs. New York has mandated retailers collect used batteries starting in June 2011. As a result of these pioneering efforts, shipping of recycled lithium-ion batteries is likely to become commonplace in the United States.[1]

Hazardous Material Transportation Codes

Lithium-ion cells (and batteries) are classified as Hazardous Materials (Hazmat)/ Dangerous Goods.[2] Thus, in the United States transport of lithium-ion cells that are "in commerce" is governed by Title 49 of the Code of Federal Regulations

---

[1] Kerchner GA, "Regulatory and Legislative Update," Proceedings, 28th International Battery Seminar and Exhibit, March 14-17, 2011, Ft. Lauderdale, FL.

[2] The terminology used is dependent upon the regulatory body: For example, the term Hazmat is used by the US DOT, while the term Dangerous Goods is used by the UN and ICAO.

C. Mikolajczak et al., *Lithium-Ion Batteries Hazard and Use Assessment*,
SpringerBriefs in Fire, DOI: 10.1007/978-1-4614-3486-3_3,
© Fire Protection Research Foundation 2011

(49 CFR), Parts 100–185. These codes are enforced by US Federal agents, usually agents of DOT, PHMSA, or the FAA. Each violation (at the time of this writing) is subject to a civil penalty of up to \$55,000,[3] and criminal penalty of up to 10 years in prison.[4] 49 CRF provides rules to determine the hazard class of a given material, and once that class has been determined, specifies requirements for transport of that material that may include requirements for testing, packaging, and labeling the material, as well as specific transport requirements (for example, limiting quantities that can be shipped by air). In the United States, Hazardous Material transport requirements are generally harmonized with the United Nations (UN) Recommendations on the Transport of Dangerous Goods, Model Regulations.

Specific requirements for lithium-ion cells are found in 49 CFR Part 173.185, "Lithium cells and batteries." and in a series of special provisions in Part 172.102 (special provisions 29, 188, 189, 190 A54, A55, A100, A101, A103, A104). Per 49 CFR and the UN Model Regulations, lithium-ion cells and batteries are considered Class 9 hazardous material (miscellaneous materials). The requirements listed in 49 CFR 173.185, include:

- "Be of a type proven to meet the requirements of each test in the UN Manual of Tests and Criteria," (UN Tests).
- Be equipped with an effective means of preventing external short circuits.

The special provisions exempt some lithium-ion cells from the shipping rules in Part 173.185. For example, they exempt shipments of small lithium-ion cells from some hazmat requirements if specific rules are followed, such as the cells meet the requirements of the UN tests, they are protected from short circuits, they are packaged in strong outer packages that are capable of withstanding a 1.2 m drop, etc. Except for a few very special cases such as the transport of prototypes for the purposes of testing, and the transport of cells or batteries for disposal or recycling, transport of any lithium-ion cells or batteries within the US (and generally internationally) requires that the cells or batteries meet the requirements of the UN tests.

The UN Recommendations on the Transport of Dangerous Goods, Model Regulations, and Manual of Tests and Criteria are primarily designed to ensure the safety of lithium-ion cells, battery packs, and batteries contained in, or packed with, equipment during transport. These regulations specify that in order to be shipped, lithium-ion cells or batteries must be able to pass a series of tests that have been selected to simulate extreme conditions that cargo may encounter. The UN Model Regulations and UN Tests have been developed by a UN subcommittee comprised of representatives from various nations. A number of regulatory bodies that reference or have adopted the UN tests include:

- United States Department of Transportation (DOT), in the Code of Federal Regulations (CFR) Title 49

---

[3] 49 CFR Part 107.329.
[4] 49 CFR Part 107.333.

- The International Air Transport Association (IATA), in the Dangerous Goods Regulations
- The International Civil Aviation Association (ICAO), in the Technical Instructions for the Safe Transport of Dangerous Goods
- The International Maritime Organization in the International Maritime Dangerous Goods List (IMDG)

In addition to requiring cells and batteries meet UN testing requirements, all of these organizations also impose specific packaging requirements for shipment such as:

- Requiring that cells or batteries be separated to prevent short circuits
- Requiring "strong outer packaging" or UN specification packaging
- Limits on the numbers of cells or batteries placed in a single package
- Specific labels for outer packages
- Hazardous material shipping training for employees engaged in packaging cells or batteries for transport

A summary of UN Testing Requirements from the 5th Revised Edition of the Manual of Tests and Criteria (Effective Jan. 1, 2011) is provided in Table 3.1. UN requirements for testing are periodically revised. At present an effort is underway to revise UN testing requirements for large format (electric vehicle) battery packs.

Table 3.1 UN transportation tests

UN 38.3.4.1	Test T.1— altitude simulation	Cells and batteries stored at a pressure of 11.6 kPa or less for at least 6 h at ambient temperature
UN 38.3.4.2	Test T.2— thermal cycling	Rapid thermal cycling between high- (75°C/167°F) and low- (−40°C/−40°F) storage temperatures
UN 38.3.4.3	Test T.3— vibration	Vibration exposure: sinusoidal waveform with a logarithmic sweep from 7 Hz (1 g peak acceleration) to 200 Hz (8 g peak acceleration) and back to 7 Hz; 12 cycles, 3 perpendicular mounting positions
UN 38.3.4.4	Test T.4— shock	Shock exposure: half-sine shock, 150 g peak acceleration, 6 ms pulse duration, three shocks in positive and negative directions for each of three perpendicular mounting positions (total of 18 shocks)
UN 38.3.4.5	Test T.5— external short circuit	Short circuit of less than 0.1 ohm at 55°C (131°F), 1 h duration
UN 38.3.4.6	Test T.6— impact	15.8 mm diameter bar placed across cell center, and a 9.1 kg mass is dropped onto the bar from 61 cm height
UN 38.3.4.7	Test T.7— overcharge	Over current (2X manufacturer's recommended maximum) and over voltage (for 18 V packs or less, charge to the lesser of 22 V or 2X recommended charge voltage. For >18 V packs, charge to 1.2X recommended charge voltage) charge (applied to battery packs only)
UN 38.3.4.8	Test T.8— forced discharge	Over-discharge cells a single time

Consumer Electronics Standards

Standards from a few key organizations have had a significant impact on lithium-ion battery development and safety. These organizations are UL, IEC, and IEEE. These standards are "consensus" standards which are developed with input from representatives from industry, user, academic, and government groups. Generally, the membership of a standards development group must vote on and approve changes to these standards. Although these standards are voluntary within the US, many of the required tests described are highly derivative of the UN testing requirements, and thus most cells in commerce will meet most of the voluntary standard requirements. In addition, all of the major US cell phone carriers require the CTIA mark, and thus, effectively make compliance with IEEE 1725 mandatory for the majority of cell phones. Similarly, major US distributors of consumer electronics devices generally have policies requiring UL listing of devices thus making compliance with UL testing requirements widespread. We discuss the relevant standards developed by these organizations below.

UL Standards

Two UL standards are particularly important for lithium-ion cells and batteries: UL 1642, "Standard for Lithium Batteries," and UL 2054, "Standard for Household and Commercial Batteries." Both of these standards are written for the purpose of ensuring consumer safety; in particular, they are designed to "reduce the risk of fire or explosion when batteries are used in a product" and to "reduce the risk of injury to persons due to fire or explosion when batteries are removed from a product to be transported, stored, or discarded."[5] User-replaceable batteries, as opposed to technician-replaceable batteries, are subject to additional test requirements, including flaming particle and projectile tests. A summary of UL testing requirements from both UL 1642 and UL 2054 is found in Table 3.2. Note that many of the tests in the two standards are identical.

UL is in the process of developing or updating a number of additional standards for lithium-ion battery applications including:

- UL Subject 1973 Batteries for use in Light Electric Rail (LER) applications and stationary applications
- UL Subject 2271 Batteries for use in Light Electric Vehicle (LEV) Applications
- UL Subject 2575 Standard for Lithium Ion Battery Systems for Use in Electric Power Tool and Motor Operated, Heating and Lighting Appliances
- UL Subject 2580 Batteries for use in electric vehicles

---

[5] UL 1642, "Lithium Batteries," 1995, p. 5; UL 2054, "Household and Commercial Batteries," 1997, p. 5.

Table 3.2 UL tests

UL 1642, Sec 10	Short-circuit test	Short circuit the cell through a maximum resistance of 0.1 ohm; testing at 20°C (68°F) and 55°C (131°F); testing of fresh and cycled cells
UL 1642, Sec 11	Abnormal charging test	Over-current charging test (constant voltage, current limited to 3X specified max charging current); testing at 20°C (68°F); testing of fresh and cycled ("conditioned") cells; 7 h duration
UL 1642, Sec 12	Forced-discharge test	For multi-cell applications only; over-discharge test; testing at 20°C (68°F); testing of fresh and cycled cells
UL 1642, Sec 13	Crush test	Cell is crushed between two flat plates to an applied force of 13 kN (3,000 lbs); testing of fresh and cycled cells
UL 1642, Sec 14	Impact test	16 mm diameter bar is placed across a cell; a 9.1 kg (20 lb) weight is dropped on to the bar from a height of 24 inches (61 cm); testing of fresh and cycled cells
UL 1642, Sec 15	Shock test	Three shocks applied with minimum average acceleration of 75 g; peak acceleration between 125 and 175 g; shocks applied to each perpendicular axis of symmetry; testing at 20°C (68°F); testing of fresh and cycled cells
UL 1642, Sec 16	Vibration test	Simple harmonic vibration applied to cells in three perpendicular directions; frequency is varied between 10 and 55 Hz; testing of fresh and cycled cells
UL 1642, Sec 17	Heating test	Cell or battery placed into an oven initially at 20°C (68°F); oven temperature is raised at a rate of 5°C/min (9°F/min) to a temperature of 130°C (266°F); the oven is held at 130°C for 10 min, then the cell is returned to room temperature; testing of fresh and cycled cells
UL 1642, Sec 18	Temperature cycling test	Cell is cycled between high- and low-temperatures: 4 h at 70°C (158°F), 2 h at 20°C (68°F), 4 h at −40°C (−40°F), return to 20°C, and repeat the cycle a further nine times; testing of fresh and cycled cells
UL 1642, Sec 19	Low pressure (altitude simulation) test	Cell is stored for 6 h at 11.6 kPa (1.68 psi); testing at 20°C (68°F); testing of fresh and cycled cells
UL 2054, Sec 9	Short-circuit test	Short circuit the cell through a maximum resistance of 0.1 ohm; testing at 20°C (68°F) and 55°C (131°F); testing of fresh cells
UL 2054, Sec 10	Abnormal charging test	Over-current charging test (constant voltage, current limited to 3X specified max charging current); testing at 20°C (68°F); testing of fresh cells
UL 2054, Sec 12	Forced-discharge test	For multi-cell applications only; over-discharge test; testing at 20°C (68°F); testing of fresh cells
UL 2054, Sec 14	Crush test	Cell is crushed between two flat plates to an applied force of 13 kN (3,000 lbs); testing at 20°C (68°F); testing of fresh cells
UL 2054, Sec 15	Impact test	15.8 mm diameter bar is placed across a cell; a 9.1 kg (20 lb) weight is dropped on to the bar from a height of 61 cm (24 inches); testing at 20°C (68°F); testing of fresh cells

(continued)

Table 3.2 (continued)

UL 2054, Sec 16	Shock test	Multiple shocks applied with minimum average acceleration of 75 g; peak acceleration between 125 and 175 g; shocks applied to each perpendicular axis of symmetry, testing at 20°C (68°F); testing of fresh cells
UL 2054, Sec 17	Vibration test	Simple harmonic vibration applied to cells in three perpendicular directions; frequency is varied between 10 and 55 Hz; testing at 20°C (68°F); testing of fresh cells
UL 2054, Sec 23	Heating test	Cell or battery placed into an oven initially at 20°C (68°F); oven temperature is raised at a rate of 5°C/min (9°F/min) to a temperature of 130°C (266°F); the oven is held at 130°C for 10 min, then the cell is returned to room temperature; testing of fresh cells
UL 2054, Sec 24	Temperature cycling test	Cell is cycled between high and low temperatures: 4 h at 70°C (158°F), 2 h at 20°C (68°F), 4 h at −40°C (−40°F), return to 20°C, and repeat the cycle a further nine times; testing of fresh cells

IEC Standards

As mentioned above, IEC Standard CEI/IEC 61960 (see footnote 33 in Chap. 1). provides a description of standard cell designations. This standard also provides procedures for assessing cell performance under a variety of conditions, such as various temperatures, various discharge rates, and after extended cell cycling. It is intended as a commodity standard that allows battery purchasers to compare performance of different cells under a single set of tests. It does not include safety tests.

The IEC publishes a standard that specifically addresses safety requirements for rechargeable cells and batteries: CEI/IEC 62133.[6] This standard is voluntary in the United States, but is required for cells and battery packs used in telecommunication devices shipped to Brazil. Per this standard, *"Cells and batteries shall be so designed and constructed that they are safe under conditions of both intended use and reasonably foreseeable misuse."* CEI/IEC 62133 includes a set of design and manufacturing requirements, as well as a series of safety tests. These requirements are summarized in Table 3.3. Many of the tests described are very similar to those described in UL and IEEE Standards (shaded portion of the table). However, a few tests specified by IEC are unique; particularly, a "free fall" test which requires multiple drops of a cell or battery onto a concrete floor, and an overcharge test for cells that requires protection from high-voltage overcharging.

IEC publishes another standard that specifically addresses safety requirements for rechargeable cells and batteries during transport: IEC 62281.[7] This standard

---

[6] CEI/IEC 62133 2002-10, "Secondary cells and batteries containing alkaline or other non-acid electrolytes—Safety requirements for portable sealed secondary cells, and for batteries made from them, for use in portable applications."

[7] IEC 62281:2004, Safety or primary and secondary lithium cells and batteries during transport.

expands on the design and manufacturing requirements specified in IEC 62133, includes the UN T-tests, adds a packaging drop test (listed in Table 3.3), and describes packaging markings. This standard is currently in revision to harmonize with changes in the UN testing requirements.

Table 3.3 IEC design requirements and safety tests

62133: 2.1	Insulation and wiring	Minimum electrical resistance requirements for positive terminal and internal wiring insulation for batteries
62133: 2.2	Venting	Requirement for a pressure relief mechanism on cells
62133: 2.3	Temperature/current management	Requirement to prevent abnormal temperature rise (by limiting charge/discharge currents)
62133: 2.4	Terminal contacts	Requirements for polarity marking, mechanical strength, current carrying capability, and corrosion resistance
62133: 2.5	Assembly of cells into batteries	Requirements for matching cells' capacity and compatibility for assembly into battery packs, and for preventing cell reversal
62133: 2.6	Quality plan	Requirement for a manufacturer quality plan
62133: 4.3.2	Short-circuit test	Short circuit the cell through a maximum resistance of 0.1 ohm; testing at 20°C (68°F) and 55°C (131°F); testing of fresh cells
62133: 4.3.11	Abnormal charging test	Over-current charging test (constant voltage, current limited to 3X specified max charging current); testing at 20°C (68°F); testing of fresh cells
62133: 4.3.10	Forced-discharge test	For multi-cell applications only; over-discharge test; testing at 20°C (68°F); testing of fresh cells
62133: 4.3.6	Crush test	Cell is crushed between two flat plates to an applied force of 13 kN (3,000 lbs); testing of fresh cells
62133: 4.3.4	Shock test	3 shocks applied with minimum average acceleration of 75 g; peak acceleration between 125 and 175 g; shocks applied to each perpendicular axis of symmetry; testing at 20°C (68°F); testing of fresh cells
62133: 4.2.2	Vibration test	Simple harmonic vibration applied to cells in three perpendicular directions; frequency is varied between 10 and 55 Hz; testing at 20°C (68°F); testing of fresh cells
62133: 4.3.5	Heating test	Cell or battery placed into an oven initially at 20°C (68°F); oven temperature is raised at a rate of 5°C/min to a temperature of 130°C (266°F); the oven is held at 130°C for 10 min, then the cell is returned to room temperature; testing of fresh cells
62133: 4.2.4	Temperature cycling test	Cell is cycled between high and low temperatures: 4 h at 75°C (167°F), 2 h at 20°C (68°F), 4 h at −20°C (−4°F), return to 20°C, and repeat the cycle a further four times; testing of fresh cells

(continued)

Table 3.3 (continued)

62133: 4.3.7	Low pressure (altitude simulation) test	Short circuit the cell through a maximum resistance of 0.1 ohm; testing at 20°C (68°F) and 55°C (131°F); testing of fresh cells
62133: 4.2.1	Continuous low-rate charging	Fully charged cells subjected to manufacturer specified charging for 28 days, testing at 20°C (68°F); testing of fresh cells
62133: 4.2.3	Moulded case stress at high ambient temperature	Battery is placed into an air-circulating oven at 70°C (158°F) for 7 h; testing of fresh batteries
62133 4.3.3	Free fall	Each cell or battery is dropped three times from a height of 1 m onto a concrete floor
62133: 4.3.9	Overcharge	A discharged cell is charged by a power supply at a minimum of 10 V for an extended period
62281: 6.6.1	Package drop test	A package filled with cells or battery packs is to be dropped from a height of 1.2 m onto a concrete surface such that one of its corners hits the ground first

IEEE Standards

In response to reported incidents of lithium-ion battery field failures, two IEEE standards underwent significant revision in the last decade: IEEE 1725 (Table 3.4),[8] "Standard for Rechargeable Batteries for Cellular Telephones," and shortly thereafter IEEE 1625,[9] "Standard for Rechargeable Batteries for Multi-Cell Mobile Computing Devices." In the US, the IEEE Standards are voluntary. However, cell phone carriers, through CTIA (The Wireless Association) have mandated compliance with IEEE 1725 to their suppliers. IEEE 1725 and 1625 share a great deal in common and will be discussed jointly. Fundamentally, both standards emphasize that battery pack safety is a function of a number of inter-related components: the cells, the battery pack, the host device, the power supply accessories, the user, and the environment. The IEEE Standards establish that the "... *responsibility for total system reliability is shared between the designers/ manufacturers/suppliers of the subsystems and the end user*" (see footnote 8). Both standards require a design analysis for a system using tools such as failure modes and effects analysis (FMEA) or fault tree analysis. The purpose of this analysis is to "... *consider all system usage scenarios ... and the associated affected subsystems*" (see footnote 8). Both standards attempt to describe and encompass industry best practices in the areas of cell, pack, system, and charging accessory design and manufacturing. Both standards require cells and battery

---

[8] IEEE 1725-2006, Standard for Rechargeable Batteries for Cellular Telephones.

[9] IEEE 1625-2008, Standard for Rechargeable Batteries for Multi-Cell Mobile Computing Devices.

Table 3.4 Unique IEEE 1625 and 1725 safety tests

1725 5.6.5 1625 5.6.6 and 5.6.7.2	Cell thermal test	Cell or battery placed into an oven initially at 20°C (68°F); oven temperature is raised at a rate of 5°C/min (9°F/min) to a temperature of 130°C (266°F); the oven is held at 130°C for 1 h, then the cell is returned to room temperature; testing of fresh cells for 1725, fresh and cycled cells for 1625
1725 5.6.6	Evaluation of excess lithium plating	Production lot of cells cycled 25 times at maximum charge/discharge rate specified by the manufacturer at 25°C (77°F). Minimum five cells then to be subjected to UL external short circuit test at 55°C (131°F). Minimum five cells dissected and examined for evidence of lithium plating
1725 5.6.7	External short circuit	Short circuit the cell through a maximum resistance of 0.05 ohm; testing at 55°C (131°F); testing of fresh cells
1725 6.14.5.1	Validation of maximum voltage protection	Cell is subjected to the maximum voltage allowed by protection electronics; cell is insulated to create adiabatic conditions for 24 h
1725.6.14.6	Pack drop test	Fully charged packs dropped from a height of 1.5 m onto smooth concrete for up to six repetitions on six sides (36 times); open circuit voltage monitored for evidence of internal shorts

packs to comply with UN and UL 1642 requirements, and recommend testing to UL 2054 and IEC 62133 requirements. The two standards include some additional testing that goes beyond the standard tests already described. In particular, both tests require that cells can withstand exposure to 130°C (266°F) conditions for 1 h (compared to the 10 min required by UN and UL tests).

Automotive Application Standards

Lithium-ion battery development in the automotive industry is in a formative stage. At the time of this writing, there are no standard cell sizes or form factors, module or pack sizes or form factors, pack voltage requirements, or protection electronics approaches. A series of test manuals have been released by the US DOE including:

- US ABC Technology Assessment Test Plan, Issued November 30, 2009
- INL/EXT-07-12536 Battery Test Manual For Plug-In Hybrid Electric Vehicles, Revision 0, Issued March 2008
- SAND 2005-3123 Freedom CAR Electrical Energy Storage System Abuse Test Manual for Electric and Hybrid Electric Vehicle Applications, Issued June 2005

- INEEL/EXT-04-01986 Battery Technology Life Verification Test Manual, Issued February 2005
- Electric Vehicle Battery Test Procedures Manual, Revision 2, Issued January 1996

Standards organizations such as IEC, the European Committee for Standardization (CEN), L'Institut National de l'Environnement Industriel et des Risques (INERIS), Japan Electric Vehicle Association (JEVA), the International Organization for Standardization (ISO), the Society of Automotive Engineers (SAE), and UL are in the process of drafting new safety and performance standards for electric and hybrid electric vehicle batteries. These standards include:

- IEC 62660-1, Work in progress, "Secondary lithium-ion cells for the propulsion of electric road vehicles—Part 1: Performance testing"
- IEC 62660-2, Work in progress, "Secondary lithium-ion cells for the propulsion of electric road vehicles—Part 2: Reliability and abuse testing"
- INERIS ELLICERT Version D, October 2010, "Certification Scheme for Battery Cells and Packs for Rechargeable Electric and Hybrid Vehicles"
- ISO 26262,[10] Work in progress, "Road vehicles—Functional Safety"
- ISO 12405, Work in progress, "Electrically propelled road vehicles—Test specification for lithium-ion traction battery packs and systems—Part 1: High power applications"
- ISO 12405, Work in progress, "Electrically propelled road vehicles—Test specification for lithium-ion traction battery packs and systems—Part 2: High energy applications"
- ISO 12405, Work in progress, "Electrically propelled road vehicles—Test specification for lithium-ion traction battery packs and systems—Part 3: Safety performance requirements"
- SAE J2929, FEB 2011, "Electric and Hybrid Vehicle Propulsion Battery System Safety Standard – Lithium-based Rechargeable Cells"
- SAE J2344, MAR 2010, "Guidelines for Electric Vehicle Safety"
- SAE J1772, JAN 2010, "SAE Electric Vehicle and Plug in Hybrid Electric Vehicle Conductive Charge Coupler"
- SAE J2464, NOV 2009, "Electric and Hybrid Electric Vehicle Rechargeable Energy Storage System (RESS) Safety and Abuse Testing"
- SAE J2380, MAR 2009, "Vibration Testing of Electric Vehicle Batteries"
- SAE J2289, JUL 2008, "Electric-Drive Battery Pack System: Functional Guidelines"
- SAE J1798, JUL 2008, "Recommended Practice for Performance Rating of Electric Vehicle Battery Modules"

---

[10] While it does not explicitly pertain to batteries, ISO 26262 is expected to have a significant impact on the system level design criteria of battery packs for hybrid electric vehicles (HEV), plug in hybrid electric vehicles (PHEV), and electric vehicles (EV).

- SAE J1797, JUN 2008, "Recommended Practice for Packaging of Electric Vehicle Battery Modules"
- SAE J2288, JUN 2008, "Life Cycle Testing of Electric Vehicle Battery Modules"
- UL Subject 2580,[11,12] Work In Progress, "The Subject Standard for Safety of Batteries for Use in Electric Vehicles"

Even though some of these standards have been recently reissued, many have not yet been updated to be applicable to lithium-ion technologies, and are certainly not mature standards. For example, in the preamble to SAE J2344, "Guidelines for Electric Vehicle Safety," which was last revised in 2010, the authors note, *"The architecture and the chemistry of EVs HV source, has also significantly changed since this document was issued due to newer technologies and packaging ... This document is being updated to include these variations and additions."*

SAE J2464, NOV 2009, provides for general hazardous and flammable material monitoring, as well as some new tests appropriate to the automotive environment. However, there are a number of indications that this standard is not yet mature:

- Much of the testing described appears derived from commercial electronics battery standards and does not appear fully adopted to automotive requirements.
- The standard includes a test described as "cell forced vent without thermal runaway," which in practice is almost impossible to achieve with traditional lithium-ion cells.
- The standard allows the tester to select a method for achieving venting with thermal runaway. It does note that the method selected may affect vent gas composition; a more mature standard would specify a series of specific abuse conditions.

Individual automakers are in the process of writing their own internal standards. The international regulatory community continues to work on appropriate requirements for transport electric and hybrid electric vehicle batteries (see footnote1). Note that there are considerable experimental challenges and expenses involved in testing large format cells and battery packs,[13] so newly developed standards will likely require testing very limited quantities of cells and battery packs (e.g., testing a single battery pack rather than multiple packs).

---

[11] http://www.ul.com/global/eng/pages/offerings/industries/powerandcontrols/electricvehicle/evstandards/

[12] "Underwriters Laboratories Working on Standards for Electric-Car Batteries," http://blogs.edmunds.com/greencaradvisor/2009/09/underwriters-laboratories-working-on-standards-for-electric-car-batteries.html.

[13] Staubel JB, "Safety Testing of Tesla's Battery Packs," Proceedings, Ninth International Advanced Automotive Battery and EC Capacitor Conference (AABC), June 2009.

Fire Protection Standards

At present, Exponent is not aware of any fire protection standards specific to lithium–ion cells. NFPA 13 *Standard for the Installation of Sprinkler Systems* currently does not provide a specific recommendation for the commodity classification (or fire protection strategies) for lithium-ion cells or complete batteries containing several cells. Exponent is not aware of any applicable International Code Council (ICC) codes that reference lithium-ion cells or battery packs. None of the widely accepted standards applicable to lithium-ion battery packs include water application tests. A further discussion of this issue is included in Chap. 7.

Chapter 4
Lithium-Ion Battery Failures

The fact that batteries can fail on rare occasions in an uncontrolled manner has brought an increased public awareness for battery safety, in particular as a result of some very large product recalls of portable notebook computer and cell phone batteries.[1]

Both energetic and non-energetic failures of lithium-ion cells and batteries can occur for a number of reasons including: poor cell design (electrochemical or mechanical), cell manufacturing flaws, external abuse of cells (thermal, mechanical, or electrical), poor battery pack design or manufacture, poor protection electronics design or manufacture, and poor charger or system design or manufacture. Thus, lithium-ion battery reliability and safety is generally considered a function of the entirety of the cell, pack, system design, and manufacture (see footnote 8, 9 in Chap. 3).

Performance standards listed in Chap. 3: Summary of Applicable Codes and Standards are designed to test cell and battery pack designs. At the time of this

---

[1] For example, see: US Consumer Products Safety Commission, Alert #10-752, "Asurion Recalls Counterfeit BlackBerry®-Branded Batteries Due to Burn and Fire Hazards," August 10, 2010.

US Consumer Products Safety Commission, Release #10-240, "HP Expands Recall of Notebook Computer Batteries Due to Fire Hazard," May 21, 2010.

US Consumer Products Safety Commission, Release #10-169, "Mobile Power Packs Recalled By Tumi Due to Fire Hazard," March 17, 2010.

US Consumer Products Safety Commission, Release #09-045, "Lithium-Ion Batteries Used with Bicycle Lights Recalled By DiNotte Lighting Due to Burn Hazard," November 18, 2008.

Darlin D, "Dell Recalls Batteries Because of Fire Threat," The New York Times, August 14, 2006.

Kelley R, "Apple recalls 1.8 million laptop batteries," CNNMoney.com, August 24, 2006: 4:38 PM EDT.

US Consumer Products Safety Commission, Release #06-231, "Dell Announces Recall of Notebook Computer Batteries Due to Fire Hazard," August 15, 2006.

US Consumer Products Safety Commission, Release #06-245, "Apple Announces Recall of Batteries Used in Previous iBook and PowerBook Computers Due to Fire Hazard," August 24, 2006.

C. Mikolajczak et al., *Lithium-Ion Batteries Hazard and Use Assessment*,
SpringerBriefs in Fire, DOI: 10.1007/978-1-4614-3486-3_4,
© Fire Protection Research Foundation 2011

writing, failures that occur in the field are seldom related to cell design, but are rather predominantly the result of manufacturing defects or subtle abuse scenarios that result in the development of latent cell internal faults. This can be considered one of the successes of the existing standards, and thus there is no strong link between failure modes actually observed in the field to the performance standards in Chap. 3: Summary of Applicable Codes and Standards. Ideally, an understanding of likely field failure mechanisms should provide direction for additional performance standards. However, at present there is no obvious performance standard that could be implemented readily to prevent field failures due to subtle manufacturing defects or abuse. The IEEE 1625 and 1725 Standards include requirements for the application of manufacturing best practices to attempt to reduce the rate of manufacturing defects. The Battery Association of Japan (BAJ) induced short circuit test[2] has been adopted by Japan but remains controversial and not applicable to large format cells because of the safety considerations associated with conducting the test (a fully charged cell is opened, a contaminant is placed within the cell electrodes, and the electrodes are then compressed to initiate a short circuit). UL is in the process of developing an Indentation Inducing Internal Short Circuit test for simulating internal faults. This test is being designed to indent a cell in such a way as to puncture separator within the cell (and induce an internal short) without puncturing the cell casing. This test method has been demonstrated on 18650 cells, but has not yet been demonstrated with prismatic or soft-pouch cells. Other organizations such as NASA and NREL have also been experimenting with internal short inducing test protocols.

In this chapter we discuss various known lithium-ion failure modes, and when during a cell or battery pack's life cycle they are most likely to occur (storage, transport prior to usage, early usage, after extended usage, during transport for disposal), as well as under what usage conditions they are most likely to occur (charging, discharging, storage, constant docking, SOC, and temperature).

Cell and Battery Failure Modes

Non-Energetic Failures

Lithium-ion batteries can fail in both non-energetic and energetic modes. Typical non-energetic failure modes (usually considered benign failures) include loss of capacity, internal impedance increase (loss of rate capability), activation of a permanent disabling mechanism such as a CID, shutdown separator, fuse, or battery pack permanent disable, electrolyte leakage with subsequent cell dry-out, and cell swelling.

---

[2] This test is contained in JIS C 8714:2007, "Safety tests for portable lithium ion secondary cells and batteries for use in portable electronic applications."

Some of these non-energetic failure modes are commonly associated with cell-aging[3] mechanisms. The ideal lithium-ion battery failure mode is a slow capacity fade and internal impedance increase caused by normal aging of the cells within the battery. If a cell exhibits this failure mode, capacity will decrease and impedance will increase until the point the battery can no longer satisfy the power requirements of the device and must be replaced. The bulk of lithium-ion batteries in the field experience this type of failure.

Other non-energetic failure modes are related to root causes that can result in either energetic or non-energetic failures, depending upon the specific conditions under which failure occurs. Whether a specific root cause (initiating fault) will result in an energetic (i.e., cell venting, ignition of cell vent gases, or rapid disassembly of the cell) or non-energetic failure depends upon whether the initiating fault can cause sufficient heating of the cell to lead to a self-sustaining exothermic reaction within the cell. For certain failure root causes, rates of non-energetic failures will be linked to rates of energetic failures (e.g., a manufacturing defect may result in a high rate of benign warranty return failures, and occasionally, a thermal runaway failure event).

Electrolyte leakage can occur as the result of mechanical damage to cells or it can result due to internal corrosion of cells. Leakage from polymer cells is more common than leakage from hard case cells. Polymer cell seals are more delicate and failures of cell pouch protective coatings can result in pouch corrosion. In small cells, there is very little free electrolyte: it is primarily absorbed by active material. Puncture of a small cell is unlikely to result in escape of more than a few droplets of electrolyte. However, in some large format cell designs, there is an appreciable amount of free, liquid electrolyte within the cell case. For these cells, a puncture could cause a spill of hazardous material. The size of the spill would be governed by the volume of electrolyte contained in a cell, the size of the puncture, and the evaporation rate of the electrolyte solvent. Electrolyte leakage poses two potential safety hazards: human contact with electrolyte and electrolyte residue, and short-circuiting of adjacent electronic systems. An increase of internal pressure within prismatic or pouch cells will cause swelling. Swelling can be caused by a variety of non-ideal chemical reactions including: overcharge, elevated temperature aging, and moisture intrusion. Cell swelling can ameliorate some failure modes making it less likely that a cell enters thermal runaway, but it can also result in enhanced cell leakage rates. Swelling commonly results in damage to battery pack enclosures.

Cell and battery pack designs often include mechanisms to permanently disable cells or batteries if their performance degrades significantly; thus, forcing a graceful failure rather than a thermal runaway reaction. For example, a number of environmental and cell manufacturing factors can result in abnormal aging of cells that result in elevated internal impedance and early capacity fade (some of these

---

[3] The chemical reactions that occur in lithium-ion cells are not all irreversible: nonreversible side reactions typically occur as slow rates resulting in cell aging.

will be discussed below). At the cell level, CIDs or thermal fuses may be activated by elevated temperatures or elevated pressures associated with increased internal impedance and permanently disable the cell. Abnormal aging of a prismatic or polymer cell may cause that cell to swell, separating the electrodes such that continued operation becomes impossible.

In multi-cell applications where cells are connected in series, individual series element voltages are measured, and charge and discharge is terminated based on the voltage of the weakest (lowest capacity, highest impedance) series element. Thus, a single abnormally aged series element (e.g., a block of cells that is exposed to higher temperatures than neighboring cells), will cause reduced capacity of the entire pack. Such battery pack behavior may force retirement of the battery pack. Alternatively, if the pack electronics design includes block imbalance detection,[4] large capacity imbalances can drive permanent disabling of the battery pack.

Energetic Failures: Thermal Runaway

Cell thermal runaway refers to rapid self-heating of a cell derived from the exothermic chemical reaction of the highly oxidizing positive electrode and the highly reducing negative electrode; it can occur with batteries of almost any chemistry. In a thermal runaway reaction, a cell rapidly releases its stored energy. The more energy a cell has stored, the more energetic a thermal runaway reaction will be. One of the reasons lithium-ion cell thermal runaway reactions can be very energetic is these cells have very high-energy densities compared to other cell chemistries. The other reason that lithium-ion cell thermal runaway reactions can be very energetic is because these cells contain flammable electrolyte, and thus, not only do they store electrical energy in the form of chemical potential energy, they store appreciable chemical energy (especially compared to cells with water-based electrolytes) in the form of combustible materials.

The likelihood of initiating cell thermal runaway is analogous to the likelihood of ignition of many typical combustion reactions: for initiation of cell thermal runaway (or ignition of fuel), the rate of heat generation must exceed the rate of heat loss. As discussed above, self-heating of lithium-ion graphitic anodes in the presence of electrolyte initiates at temperatures in the 70–90°C (158–194°F) range. Thus, if a cell is brought to this initiating temperature in an adiabatic environment, it will eventually self-heat to the point thermal runaway initiates. For a typical 100% SOC 18650 cell brought to its self-heating temperature, thermal runaway will occur after approximately two days if the cell is well-insulated. Should initial temperature be higher, time to thermal runaway will be shorter. For example, if a typical lithium-ion cell is placed into an oven at more than 150°C (300°F), such

---

[4] For an example of battery pack protection electronics with imbalance detection, see http://focus.ti.com/docs/prod/folders/print/bq29330.html.

that separator melting occurs, additional heating due to shorting between electrodes will occur and cell thermal runaway will initiate within minutes. However, if heat is allowed to escape, time to thermal runaway may be longer, or the cell may never achieve thermal runaway. UN and UL consumer electronics standards (discussed in Chap. 3: Summary of Applicable Codes and Standards) effectively govern the minimum thermal stability of cells: they require that fully charged cells withstand extended storage at 70 or 75°C (158 or 167°F; 4 h or more), and short exposure (10 min) to 130°C (266°F) conditions. IEEE Standards require storage at 130°C for 1 h.

The severity of a cell thermal runaway event will depend upon a number of factors including the SOC of a cell (how much electrical energy is stored in the form of chemical potential energy), the ambient environmental temperature, the electrochemical design of the cell (cell chemistry), and the mechanical design of the cell (cell size, electrolyte volume, etc.). More details about the effects of these factors will be discussed below. For any given cell, the most severe thermal runaway reaction will be achieved when that cell is at 100% SOC, or is over-charged, because the cell will contain maximum electrical energy. If a typical fully charged (or overcharged), lithium-ion cell undergoes a thermal runaway reaction a number of things occur.

1. *Cell internal temperature increases.* Exponent and others have measured cell case temperatures during thermal runaway reactions. For fully charged cells, these temperatures can reach in excess of 600°C (1,110°F); case temperatures for lithium-iron phosphate cells are generally lower. The temperature rise is driven by reactions of the electrodes with electrolyte and release of stored energy. Some cathode materials will decompose and may change their crystalline structure. This structural change may result in the release of small quantities of oxygen that can participate in reactions internal to the cell (e.g., oxidation of the aluminum current collector). This fact has led to a misconception that lithium-ion cells burn vigorously because they "produce their own oxygen." This idea is incorrect. No significant amount of oxygen is found in cell vent gases.[5] Any internal production of oxygen will affect cell internal reactivity (see footnote 11 in Chap. 1), cell internal temperature, and cell case temperature, but plays no measurable role in the flammability of vent gases. Internal temperature increase results in separator melting and decomposition, and usually, melting of the aluminum current collector, which occurs at 660°C (1,220°F). Liquid aluminum may alloy with any exposed copper within the cell. Some copper and aluminum alloys have melting points as low as 548°C (1,018°F), so damage to the internal copper current collectors is likely to occur.

---

[5] Analysis of cell headspace gases can reveal the presence of argon, nitrogen, and oxygen consistent with cell construction conditions. In one instance (testing of a prototype cell), trace quantities of oxygen and hydrogen were measured in cell vent gases, but spark ignition testing of those gases did not result in ignition. See Roth EP, Crafts CC, Doughty DH, McBreen J, "Thermal Abuse Performance of 18650 Li-Ion Cells," Sandia Report SAND2004-0584, March 2004.

Fig. 4.1 An 18650 cell that has undergone thermal runaway

Fig. 4.2 An 18650 cell after
thermal runaway—
resolidified beads of melted
aluminum are visible

Temperatures produced by cell thermal runaway reactions are considered sufficient to cause hot surface ignition of flammable mixtures, but do not reach
levels that will cause the melting of pure copper (1,080°C/1,976°F), nickel, or
steel.[6] Figures 4.1, 4.2, 4.3, 4.4 show an 18650 cell that has undergone thermal
runaway: aluminum within the cell melted, the cell separator consumed. What
remains is the cell case (steel), the copper current collector from the anode, and
a black friable material composed primarily of cathode material.

2. *Cell internal pressure increases.* This occurs because heated electrolyte will
both vaporize and decompose, and some cathode materials can also decompose,
releasing gas. In a pouch or prismatic cell, this will result in cell swelling. For a
typical cylindrical design, appreciable swelling will not occur. However, if a

---

[6] Sometimes evidence of very small points of pure copper, nickel, or steel melting are found
within a cell. These points are the result of internal electric arcing/shorting and are not indicative
of overall cell thermal runaway temperatures.

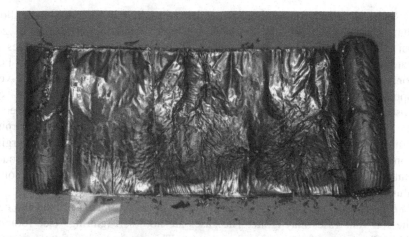

Fig. 4.3 Unrolling windings of an 18650 cell that underwent thermal runaway—note the copper current collector remains largely intact

Fig. 4.4 Internal contents of an 18650 cell that underwent thermal runaway: copper current collector (*top*) and remains of active materials (*bottom*)

cylindrical cell has been sufficiently heated (usually from an external source), the case walls may soften sufficiently to allow bulging of the cell base.

3. *Cell undergoes venting.* In a soft-pack polymer cell, the heat seals fail at fairly low temperatures, resulting in low-pressure venting. There may be an audible pop (sound) when the pouch is breached. Prismatic cell cases may have a vent port installed (large format cells) or may incorporate score marks in them to provide a weak point for case venting. In some cases, designers may have determined that the case weld points will break at appropriate cell internal pressures (e.g., small prismatic cells). Venting of small prismatic cells is usually accompanied by a loud pop. In small single cell applications (e.g., cell phones) venting usually causes the cell to eject from the device. A typical consumer description of cell thermal runaway from a cell phone is the user hears a loud sound, and upon investigation finds, the cell became detached from the device. This often creates a char mark on the surface below the cell.

Cylindrical cells have venting mechanisms installed in their cap assemblies that activate when internal pressures are high (commonly in excess of 200 psi). In most small commercial cells, CIDs connect to vent assemblies so venting is a two-stage process. First, the CID activates, creating a small hole for venting and a soft pop or click sound. Shortly thereafter, the full vent opens with a loud pop, followed by a rushing sound of venting gas. The vent gases usually appear as dark smoke. Sometimes bright sparks are observed in the vent gases.[7] Some observers have assumed these sparks are "burning lithium." However, this is highly unlikely as even under cell charging conditions, only very small quantities of lithium can plate onto electrodes. Rather, any observed bright sparks are most likely droplets of molten aluminum being ejected from the cell.

4. *Cell vent gases may ignite.* Depending upon the environment around the cell, the cell vent gases may ignite. The gases are not "self-igniting."(see footnote 27 in Chap. 1 and footnote 7 in this chapter) There must be sufficient oxygen in the surrounding environment to sustain combustion of hydrocarbons and there must be a competent ignition source to ignite the vent gases. A hot cell case could result in ignition of vent gases, as could hot metal sparks ejected with the vent gases. Lithium iron phosphate cells (cells with a $LiFePO_4$ cathode) are often described as "safer" than typical lithium cobalt oxide cells, because typical case temperatures of these cells during thermal runaway are unlikely to cause hot surface ignition of the vent gases. However, if other competent ignition sources are present, vent gases from iron phosphate cells will ignite.[8] Venting of isolated small cells (cell phone cells and smaller) seldom results in flame ignition. This is likely due to the limited volumes of vent gases released from these cells—that is, the gases become diluted before ignition can occur. In comparison, ignition of vent gases from 18650 and larger cells is fairly common: these cells contain more electrolyte (more fuel), and are usually used in multi-cell battery packs. If the flow of vent gases is "restricted" due to the configuration of a vent port (typical in hard case cells), flames emanating from the cell will be highly directional (e.g., flames from 18650 cells are often described as "torch-like").

5. *Cell contents may be ejected.* With hard case cells, internal pressure will develop prior to venting. Depending upon the mechanical design of the cell, release of pressure (due to a vent activating), may result in the ejection of the cell windings. This phenomenon is very common with cylindrical cells, particularly those without stiff center tubes (Figs. 4.5, 4.6, 4.7). Cylindrical cells have wound designs with an open center, similar to a roll of wrapping paper. When heating occurs, electrodes expand and collapse into the core of the roll; thus, creating an

[7] Webster H, "Flammability Assessment of Bulk-Packed, Rechargeable Lithium-Ion Cells in Transport Category Aircraft," DOT/FAA/AR-06/38, September 2006, http://www.fire.tc.faa.gov/pdf/06-38.pdf.

[8] Roth EP, "Abuse Tolerance Improvement," DOE Vehicle Technologies Peer Review, Gaithersburg, MD, February 26, 2008.

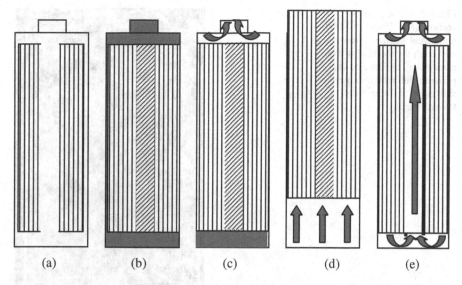

(a)   (b)   (c)   (d)   (e)

Fig. 4.5 Ejection of windings from a cylindrical cell subject to a thermal runaway reaction (*left to right*): **a** cross section of a cell without a stiff center tube; **b** during cell thermal runaway windings expand and collapse into the central core; **c** cell venting allows relief of pressure at cell cap but not at cell base; and **d** pressure at cell base acts like a piston, ejecting cell windings. In contrast, a stiff center tube **e** will maintain an open cell core and allow pressure equalization, preventing winding ejection

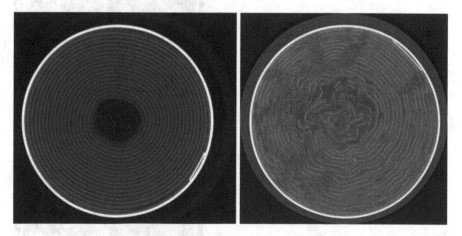

Fig. 4.6 CT scan of a normal 18650 cell showing an open center core (*left*), and a post thermal runaway 18650 cell exhibiting winding collapse into the core region (*right*)

internal obstruction to gas release at the base of the cell. When the cell vents at the cap, the pressure differential between the cell cap area and cell base can result in a piston-like effect that drives the electrodes out of the cell. Electrodes ejected in this manner (or even full cells if not well constrained by battery packs)

Fig. 4.7 CT scan cross section of an 18650 cell that underwent thermal runaway. Although the electrodes were not ejected, the base to cap pressure differential caused shifting of the electrodes toward the cap

can travel significant distances (many meters), spreading heated material, and possibly flames, far from the original battery pack.

Circa 2004–2005, US notebook computer manufactures worked with cell manufacturers to reduce the likelihood of electrode ejection in the case of cell thermal runaway. Since that time the IEEE Standards (see footnote 8, 9 in Chap. 3) have been revised and they now require cell design elements to prevent ejection of cell electrodes if thermal runaway occurs. In response to these requirements cell manufacturers began to include a stiff center tube in cylindrical cells, usually a rolled metal pin to maintain an open cell core and allow equalization of pressures within the cell in the event of a cell thermal runaway. The use of center tubes has significantly reduced the likelihood of electrode ejection on thermal runaway with cylindrical cells. However, center tubes are not a required design element for cylindrical cells and are not present in all cell designs. In addition, center tubes will not always prevent electrode ejection.

6. *Cell thermal runaway may propagate to adjacent cells.* If one cell in a pack undergoes a thermal runaway reaction, it is likely to cause thermal runaway in adjacent cells by way of various heat transfer mechanisms: direct case-to-case contact, impingement of hot vent gases, or impingement of flaming vent gases. Recent FAA tests provide a demonstration of thermal runaway propagation through bulk-packaged 18650 cells.[9] Multi-cell pack design can affect the likelihood of thermal runaway propagation by adjusting cell spacing and orientation to minimize heat transfer between adjacent cells,[10] to direct vent gases away from adjacent cells, or to increase cell cooling.

Root Causes of Energetic Cell and Battery Failures

There are a number of ways to exceed the thermal stability limits of a lithium-ion cell and cause an energetic failure. Energetic lithium-ion battery failures may be induced by external forces such as exposure to fire or severe mechanical damage, or they may be the result of problems involving charge, discharge, and/or battery protection circuitry design and implementation, or they may be caused by internal cell faults that result from rare and/or subtle manufacturing problems. Generally, the root causes of energetic cell and battery failures can be classified into:

[9] Webster H, "Fire Protection for the Shipment of Lithium Batteries in Aircraft Cargo Compartments," DOT/FAA/AR-10/31, November 2010, http://www.fire.tc.faa.gov/pdf/10-31.pdf.
[10] See for example, Spotnitz RM, Weaver J, Yeduvaka G, Doughty DH, Roth EP, "Simulation of abuse tolerance of lithium-ion battery packs," Journal of Power Sources, 163 (2007), pp. 1080–1086.

- Thermal abuse;
- Mechanical abuse;
- Electrical abuse;
- Poor cell electrochemical design; and
- Internal cell faults associated with cell manufacturing defects.

Thermal Abuse

The most direct way to exceed the thermal stability limits of a lithium-ion cell is to subject it to external heating. The external heat could be applied to the bulk of the cell, often simulated by "hot box" testing, or it could be localized to one portion of the cell, causing local reactions that propagate to the entire cell. At Exponent, we commonly use small heaters applied to the exterior of cells to initiate thermal runaway reactions and rely on bulk heat transfer to cause propagation of thermal runaway to adjacent cells in a battery pack.[11]

In Exponent's experience, very few[12] energetic field failures of consumer electronic devices have been attributed to long-term storage of cells at temperatures just above the self-heating point of 70–90°C (158–194°F). Such failures require not only elevated temperature, but an adiabatic (highly insulated) environment, and extended times to reach a self-sustaining thermal runaway condition. Although possible, these sorts of conditions are seldom achieved with consumer electronic devices in the field. They may become more likely for very dense packed large format batteries, where the high density of cells may prevent heat removal from cells at the center of the battery pack and allow long-term self-heating. Failure via this mode may also occur under certain extreme storage scenarios. Some examples might include lithium-ion batteries stored on high racks in non-climate controlled warehouses during summer months, or lithium-ion batteries stored adjacent to heaters.

Acute exposure of a cell to high temperatures (e.g., due to flame attack, exposure to hot combustion gases from a proximate fire, or contact with adjacent cells undergoing thermal runaway reactions) will readily induce thermal runaway in that cell. Typically, if an internal cell fault is sufficient to cause thermal runaway in a single cell of a multi-cell battery pack, heat transfer from the faulting cell will cause thermal runaway in neighboring cells of the battery pack. Thus, the thermal

---

[11] Harmon J, Gopalakrishnan P, Mikolajczak C, "US FAA-style flammability assessment of lithium-ion batteries packed with and contained in equipment (UN3481)," US Government Docket ID: PHMSA-2009-00095-0117, PHMSA-2009-00095-0119.1, PHMSA-2009-00095-0119.2, and PHMSA-2009-00095-0120.1, March 2010.

[12] The authors have investigated hundreds of thermal runaway failures from the field and can only ascribe one or two of the investigated failures to this failure mode.

runaway reaction will propagate through a battery pack. For example, an internal cell fault in one cell of a notebook computer battery pack will first result in thermal runaway of the faulting cell, and can subsequently cause thermal runaway reactions to propagate through all the rest of the cells in the pack. Occasionally, if heat transfer is limited between cells (e.g., the cells are well separated) thermal runaway does not propagate.

Propagation of cell thermal runaway has significant implications for fire suppression and fire protection. A fire suppressant or low oxygen environment may extinguish flames from a battery pack, but the thermal runaway reaction will propagate if heat is not sufficiently removed from the adjacent cells. Responders to fires involving lithium-ion battery packs have often described a series of re-ignition events. Typically, responders report they used a fire extinguisher on a battery pack fire, thought they had extinguished the fire, and then observed the fire re-ignite as an additional cell vented.

From a fire protection standpoint, particularly in bulk storage areas, isolation (thermal separation) of lithium-ion batteries from each other, from hot combustion products, and from oxidizers is important in mitigating and preventing fire spread following an initiating incident such a single cell undergoing thermal runaway.

Mechanical Abuse

Mechanical abuse of cells can cause shorting between cell electrodes, leading to localized cell heating that propagates to the entire cell and initiates thermal runaway. The mechanical abuse can be severe and result in immediate failure, or it can be subtle, and create a flaw in the cell that results in an internal cell fault much later (i.e., after the cell has undergone numerous cycles). UN and UL consumer electronics standards govern the minimum tolerance of cells to some forms of severe mechanical abuse: they require fully charged cells (100% SOC) withstand flat plate crushes and more concentrated point crushes perpendicular to their electrode surfaces. Some cell manufacturers will also conduct nail penetration tests and UL is considering adding a blunt nail test to UL 1642. Like crush tests, nail tests are conducted to perpendicularly penetrate electrode surfaces. The UN and UL Standards only address a few mechanical damage modes. For example, they do not address mechanical damage to electrode edges, mechanical damage to internal tab regions, or continued usage of a cell after mechanical damage has occurred.

Mechanical damage (crush or penetration) that occurs at electrode edges is significantly more likely to cause cell thermal runaway than damage perpendicular to electrode surfaces (Fig. 4.8). Exponent demonstrated this susceptibility by conducting crush tests on cells in differing orientations. When crush damage is perpendicular to electrode surfaces, it may deform the electrodes and separator layers, but it may not cause penetration of the separator, and thus, minimal or no internal shorting occurs (certainly, if the separator is penetrated, shorting and thermal runaway can occur). If the cell case is penetrated (e.g., during a nail test),

Fig. 4.8 Crush or
penetration perpendicular to
electrode edges (*red arrows*)
is more likely to cause cell
thermal runaway than crush
or penetration perpendicular
to electrode surfaces (*green
arrows*)

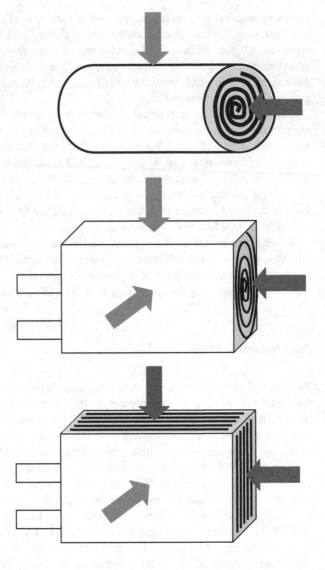

it is likely that low impedance shorting will occur between current collectors
bridged by the penetrating nail, and cell heating may be too low to result in cell
thermal runaway. However, if crush or penetration occurs perpendicular to electrode edges, that deformation is likely to result in high impedance shorting
between electrode layers and initiate cell thermal runaway. Maleki and Howard[13]
have also studied the effect of nail penetration and crush location on inducing

---

[13] Maleki H, Howard JN, "Internal short circuit in Li-ion cells," Journal of Power Sources, 191
(2009), pp. 568–574.

immediate cell thermal runaway. They found "pinch" damage at the edge of electrodes in a prismatic cell was more likely to induce immediate thermal runaway than crush damage on the flat face of a cell.

Cell susceptibility to severe mechanical damage is a factor in cell shipping and handling. Damage during handling can occur readily in a number of ways. Packages of cells, battery packs, or equipment containing packs can be subjected to severe shocks (i.e., they can be dropped), crushes, and punctures[14] causing mechanical damage to cells. This susceptibility drives the transport packaging requirements for sturdy containers. It drives the "Lithium and Lithium Ion Battery Industries' Voluntary Air Transportation Communications Program," administered by the Rechargeable Battery Association (PRBA).[15] Under this program, battery shippers mark packages containing lithium-ion cells and battery packs with warnings to quarantine packages that might be "... crushed, punctured or torn open to reveal contents." Similarly, it motivates the IEC 62281 requirement to quarantine damaged packages until contents can be inspected and repackaged (see footnote 7 in Chap. 3).

If mechanical damage does not cause cell thermal runaway immediately or within hours of occurrence, it can still cause cell thermal runaway if the cell continues to be cycled and used. A point of mild mechanical damage can become a point of electrode or separator degradation over multiple cell cycles. Ultimately, severe lithium plating (another root cause that may have the potential to result in cell thermal runaway) occurs at the point of degradation, or a significant hole in the cell separator develops, so that during or after cell charging, the cell undergoes a thermal runaway reaction. Failure by this mode, like most cell internal shorting failures, is most likely to occur during cell charging, or immediately after charging (mechanism discussed below).

Failures due to latent mechanical damage have prompted certain precautions taken by electric model aircraft (i.e., radio-controlled aircraft) enthusiasts. Lithium-ion pouch cells have long been favored by this community due to their light weight. However, because these cells are not enclosed by sturdy cases, they are very susceptible to mild mechanical damage, and routinely exposed to potential sources of mechanical damage. Fires during charging of these cells are very common even with cells that do not appear significantly damaged.[16,17] Thus, enthusiasts have learned to expect that fires will occur periodically and recommend charging cells in fire-safe areas or containers such as fire safes, fire places, and sand pits.

---

[14] In particular, forklift operations can result in puncture of packages with forklift tines. A 1999 incident at LAX involving lithium primary (lithium metal) batteries was the result of a pallet tipping over during transport via forklift and subsequent puncture of packages with the forklift tines in an attempt by the forklift operator to right the pallet.

[15] http://www.prba.org/prba/programs/lithium_ion_program/Default.ashx

[16] Schleicher R, *How to Build and Fly Electric Model Aircraft*, MBI Publishing Company, St. Paul, MN, 2005.

[17] McPherson J, Complete Guide to Lithium Polymer Batteries and LiPo Failure Reports, http://www.rcgroups.com/forums/showthread.php?t=209187.

Because of the risk associated with latent mechanical damage, battery pack assemblers, particularly those that are experienced with soft-pouch lithium-ion batteries, typically have policies in place to scrap any cell that may have become mechanically damaged, even if damage is not apparent (e.g., a cell is dropped on the floor during assembly).

Exponent has observed numerous field failures caused by latent mechanical damage, particularly of soft-pouch cells where mild mechanical damage did not cause immediate failure, but rather failed during subsequent cycling. Exponent has conducted testing to attempt to determine whether specific levels of mechanical damage will ultimately result in cell thermal runaway reactions. We have found no nondestructive way to definitely rule out a future cell failure. For example, we have X-rayed mechanically damaged cells to determine if any gross electrode deformation has occurred. However, this technique is typically not sufficient to show small cracks or delamination in electrode materials or mild amounts of electrode over-compression that can lead to lithium plating and cell thermal runaway. We have applied some destructive methods to assess whether specific levels of damage may lead to lithium-plating and cell thermal runaway. These techniques provide results specific to cell design and degree of mechanical deformation, and thus are not applicable to every lithium-ion cell.

To prevent fires from occurring due to cell mechanical damage, it is important to quarantine and monitor cells or packs that have suffered mechanical damage. Mechanically damaged cells and battery packs should then be disposed of, rather than placed back into service, unless extensive studies have been carried out specific to the cell chemistry in use and the degree of mechanical damage experienced, to show that the damage did not induce a defect likely to cause cell thermal runaway. Should mechanical damage exposure be suspected but not confirmed, the suspect batteries should be quarantined, during and after, all subsequent charging processes. Packs should be monitored for excessive self-discharge rates and charging processes should be monitored carefully for evidence of cell internal shorting (noisy voltage signals, extended charging times).[18] Any evidence of poor behavior should trigger proper disposal of the battery pack.

Electrical Abuse

There are a number of ways in which lithium-ion cells can be abused electrically, leading to cell thermal runaway reactions. Some of these mechanisms are described below.

---

[18] Mikolajczak C, Harmon J, White K, Horn Q, Wu M, Shah K, "Detecting lithium-ion cell internal fault development in real time," Power Electronics Technology; March 2010.

Overcharge

Overcharge of a lithium-ion cell can cause significant degradation of both anode and cathode. On the anode, overcharge can cause plating rather than intercalation of lithium. Plated lithium forms dendrites that can grow over time and then cause internal shorting. Plated lithium also interacts exothermically with electrolyte. On the cathode, overcharge can cause excess removal of lithium from cathode material structures, such that their crystalline structure becomes unstable, resulting in an exothermic reaction. Reactions at both the anode and cathode, as well as lithium dendrite shorting can push a cell out of its thermal stability limits and result in a thermal runaway reaction.[19] The more severe the degree of overcharge, the more likely the cell is to experience thermal runaway.

There are a few ways in which overcharge can occur. The most obvious overcharge mode is charging a cell to too high of a voltage (over voltage overcharge). For example, charging a 4.2 V rated cell above 5 V will likely cause an immediate, energetic failure. Charging at excessive currents, but not excessive voltages, can also cause an overcharge failure; in this case, localized regions of high current density within a cell will become overcharged, while other regions within the cell will remain within appropriate voltage limits.

Severe overcharge failures are not common with mature consumer electronics devices since these usually contain redundant overcharge protection mechanisms within the pack protection electronics. Occasionally, a design or manufacturing defect can cause bypassing of protection mechanisms and result in severe overcharge failures. These types of failures also occur as a result of human error with systems that either lack hardwired protection (e.g., prototype systems that are being tested) or in charging schemes with manual voltage and current settings (e.g., radio-control aircraft batteries).

Although severe overcharge will lead to immediate cell thermal runaway, repeated slight overcharge of a cell may not cause a failure for an extended timeframe, but can eventually result in thermal runaway. Until circa 2008, it was common to set secondary over voltage protection limits in multi-series battery packs to voltages that represented a slight overcharge level. As the cells in these packs aged, the capacity of series elements diverged. Then, weak cells were allowed to repeatedly reach the secondary protection limit (when a weak cell hit this voltage limit charging would terminate). This could cause repeated, slight overcharging of the weakest cells in the pack, but not the other cells. In some instances, this led to thermal runaway reactions. Industry awareness of this problem prompted requirements in IEEE 1725 and IEEE 1625 for cell manufacturers to communicate specific high voltage limits appropriate for secondary protection settings specific to each cell design to pack and device designers who

---

[19] For a detailed discussion of reactions that can occur during overcharge, see: Belov D, Yang MH, "Failure mechanism of Li-ion battery at overcharge conditions," Journal of Solid State Electrochemistry, **12** (2008), pp. 885–894.

purchase their cells. IEEE 1625 adopted the concept of a safe charging current and charging voltage envelope relative to temperature from the Battery Association of Japan (BAJ) "Guidance for Safe Usage of Portable Lithium-Ion Rechargeable Battery Pack."[20]

External Short Circuit

High rate discharging (or charging) can cause resistive heating within cells at points of high impedance. Such internal heating could cause cells to exceed thermal stability limits. Points of high impedance could include weld points within a cell (internal tab attachment) or electrode surfaces. As cell size and capacity increases, the likelihood of internal impedance heating leading to thermal runaway also increases. Larger cells exhibit slower heat transfer to their exteriors, and they usually have higher capacities. Thus, they have the potential to convert more electrical energy to internal heat. UN and UL testing requirements provide a minimum requirement for cell external short circuit resistance: discharge through a resistance of less than 0.1 ohm in a 55°C (131°F) environment. International and domestic shipping regulations (as found in the US CFR, as well as IATA and ICAO publications) require that cells or batteries be protected from short-circuiting. Investigation of a number of thermal runaway failures that have occurred during transport has revealed that improper packaging, particularly a failure to prevent short circuits is a common cause of these incidents.

Over-Discharge

Simply over-discharging a lithium-ion cell to 0 V will not cause a thermal runaway reaction. However, such over-discharge can cause internal damage to electrodes and current collectors (i.e., dissolution of copper) (Fig. 4.9), can lead to lithium plating if the cell is recharged (particularly, if the cell is repeatedly over-discharged), and can ultimately lead to thermal runaway. Most consumer electronics devices set specific discharge voltage limits for their lithium-ion battery packs, at which point an electrical switch will disconnect the electrical load from the battery pack to prevent over-discharge. This switch is reset upon charging. However, such a mechanism cannot completely prevent over-discharge. For example, a battery pack may be discharged to the low voltage cutoff and then stored for an extended period of time during which self-discharge of the cell ultimately results in over-discharge. Most pack protection electronics will allow

---

[20] "Guidance for Safe Usage of Portable Lithium-Ion Rechargeable Battery Pack," 1st Edition, March 2003, Battery Association of Japan.

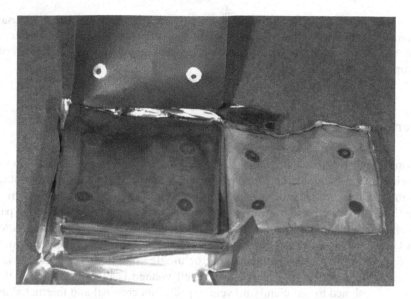

Fig. 4.9 The anode of a repeatedly over-discharged cell coated with copper

the recharge of over-discharged cells, despite the potential for the negative electrode to become damaged. Therefore, over-discharge does periodically cause thermal runaway of lithium-ion cells.

Forcing a cell into "reversal" (charging to a negative voltage, "forced over-discharge") may cause thermal runaway. UL and UN tests provide a minimum requirement for resistance to forced over-discharge for cells used in multi-cell packs. These tests are designed to simulate the most likely mechanism of forced discharge, which occurs when a cell with lower capacity than its neighboring series elements is present in a multi-series battery pack that is externally short-circuited. A lower capacity cell of this type can occur due to aging of the battery pack. In this scenario, current flow from the higher capacity series elements in the pack will drive the discharged series element into reversal. The UN and UL testing does not include repeated forced discharge. Thus, if a system does not include protection electronics that will detect and disable charging of a damaged cell, it is possible a cell could be repeatedly force over-discharged and ultimately undergo a thermal runaway reaction.

Poor Cell Electrochemical Design

Commercial cell testing will generally ensure cells perform adequately when new. However, on occasion, cell aging will result in unexpected degradation of a cell component such as one of the electrodes, the separator, or the electrolyte that can result in thermal runaway failures.[21] Usually, in these instances, field usage conditions were not well understood when the cell was designed and selected. Thus, conditions used for initial safety and reliability testing were not wholly appropriate to the application.

Internal Cell Fault Related to Manufacturing Defects

Exponent has worked on understanding cell faults for over a decade. In our experience, most commercial electronics cells and battery packs are robustly designed and do not have obvious design problems. The cell designs pass transportation and commercial testing standards and a wide range of cell and pack manufacturer internal reliability tests. Commercially available notebook battery packs have redundant protection devices in place in order to prevent cell overcharging and other potentially damaging or unsafe conditions (charging at high temperatures, charging at high rate when cell voltage is low, etc.). The battery packs are designed to sufficiently prevent expected mechanical and thermal abuse; nonetheless, thermal runaway failures still occur. In Exponent's experience, for commercial lithium-ion battery packs with mature protection electronics packages, the majority of thermal runaway failures in the field are caused by internal cell faults related to cell manufacturing defects.

There are numerous flaws that can occur during cell manufacturing that can ultimately result in cell thermal runaway reactions. Fundamentally, problems at any step of the cell manufacturing process can result in an internal cell fault. For example, there can be defects in cell raw materials, defects in electrode coatings, contaminants introduced during assembly processes, and misplaced, misapplied, or damaged components. Exponent has observed cell thermal runaway failures resulting from cell contamination (either by materials foreign to the battery or loose pieces of battery material itself), manufacturing-induced electrode damage (scratches, punctures, tears, active material displacement), burrs on electrode tabs, weld spatter from cell tab attachment points, wrinkles or kinks in windings or tabs, and electrode misalignment (Fig. 4.10).

There are many ways to classify manufacturing faults. Fouchard and Lechner[22] classified internal shorts by impedance: hard shorts were characterized by low

---

[21] See for example: Horn QC, "Application of microscopic characterization techniques for failure analysis of battery systems," Invited presentation, San Francisco Section of the Electrochemical Society, March 27, 2008.

Horn QC, White KC, "Novel imaging techniques for understanding degradation mechanisms in lithium-ion batteries," Advanced Automotive Battery Conference, Tampa, FL, May 13, 2008.

Horn Q, White KC, "Understanding lithium-ion degradation and failure mechanisms by cross-section analysis," 211th Electrochemical Society Meeting, Chicago, IL, Spring 2007.

[22] Fouchard D, Lechner L, "Analysis of Safety and Reliability in Secondary Lithium Batteries," Electrochimica Acta, **38**(9), pp. 1193–1198, 1993.

Fig. 4.10 Examples of manufacturing flaws that can lead to cell internal shorts including contamination, poor welds, weld spatter, flaws in electrode coatings, and tears in electrodes and separators

impedance resulting in quick discharge of a cell, while soft shorts were characterized by high impedance resulting in relatively slow discharge of a cell that might appear as a high self-discharge rate. They noted external protective devices could not protect against sudden hard shorts, while imbalance detection might detect soft shorts, and potentially disable battery packs preventing soft shorts from evolving into hard shorts. Exponent attempted classification by component or manufacturing process,[23] which proved applicable to cell manufacturing auditing.[24] The IEEE 1625 and 1725 Standards take this approach and include component-by-component best manufacturing practice guidance. During revisions of the IEEE 1625 Standard, efforts were made to determine critical locations within a cell where shorting is most likely to lead to heat generation.[25] This analysis led to improvements in cell design to eliminate particularly susceptible locations from commercial cell designs. Efforts have been made to characterize the "size" of a given internal short that will lead to thermal runaway ("size" can refer to the physical size of a contaminant particle, or to the amount of energy that must be transferred)[26] in order to identify mechanisms to detect incipient cell faults (see footnote 18).[27]

It is beyond the scope of this document to describe the possible cell manufacturing flaws.[28] However, a discussion of how internal faults related to

---

[23] Mikolajczak CJ, "Causes of Li-Ion Internal Cell Faults," IEEE 1625 Meeting, San Jose, CA, November 15, 2006.

Mikolajczak CJ, "Causes of Li-ion internal cell faults," Portable Rechargeable Battery Association Membership Meeting, Dallas, TX, October 12, 2006.

Mikolajczak CJ, Hayes T, Megerle MV, Wu M, "Li-Ion internal cell faults," Extended Battery Life Working Group Meeting, San Jose, CA, October 4, 2006.

[24] Hayes T, Mikolajczak C, Horn Q, "Key manufacturing practices and techniques to achieve high quality Li-ion cells," Proceedings, 27th International Battery Seminar & Exhibit for Primary & Secondary Batteries, Small Fuel Cells, and Other Technologies, Ft. Lauderdale, FL, March 15–18, 2010.

[25] Zhang Z, "Li-ion in EDV and Safety Perspectives," Proceedings, 28th International Battery Seminar & Exhibit, March 14–17, 2011, Ft. Lauderdale, FL.

[26] See for example: Mikolajczak C, Harmon J, Hayes T, Megerle M, White K, Horn Q, Wu M, "Lithium-ion battery cell failure analysis: The significance of surviving features on copper current collectors in cells that have experienced thermal runaway," Proceedings, 25th International Battery Seminar & Exhibit for Primary & Secondary Batteries, Small Fuel Cells, and Other Technologies, Ft. Lauderdale, FL, March 17–20, 2008.Barnett B, Sriramulu S, "New Safety Technologies for Lithium-Ion Batteries," Proceedings, 28th International Battery Seminar & Exhibit, March 14–17, 2011, Ft. Lauderdale, FL.

[27] Mikolajczak C, Harmon J, Stewart S, Arora A, Horn Q, White K, Wu M, "Mechanisms of latent internal cell fault formation: Screening and real time detection approaches," Proceedings, Space Power Workshop, Manhattan Beach, CA, April 20–23, 2009.

[28] Exponent has produced numerous publications on the topic of cell internal faults, including: Godithi R, Mikolajczak C, Harmon J, Wu M, "Lithium-ion cell screening: Nondestructive and destructive physical examination," NASA Aerospace Workshop, Huntsville, AL, November 2009.

Mikolajczak C, Harmon J, Wu M, "Lithium plating in commercial lithium-ion cells: Observations and analysis of causes," Proceedings, Batteries 2009 the International Power Supply Conference and Exhibition, French Riviera, September 30–October 2, 2009.

Hayes T, Mikolajczak C, Megerle M, Wu M, Gupta S, Halleck P, "Use of CT scanning for defect detection in lithium-ion batteries," Proceedings, 26th International Battery Seminar & Exhibit for Primary & Secondary Batteries, Small Fuel Cells, and Other Technologies, Ft. Lauderdale, FL, March 16–19, 2009.

Horn QC, "Battery involvement in fires: cause or effect?" Invited seminar, International Association of Arson Investigators—Massachusetts Chapter, Auburn, MA, March 19, 2009.

manufacturing defects manifest themselves can be useful from a fire protection standpoint.

An internal cell fault results in a short circuit inside a cell. If the point of shorting is minor (a micro-short), separator shutdown (i.e., physical blockage of lithium ion transport at a localized region within the cell) may isolate the flaw and allow the cell to continue functioning normally. If the point of shorting releases sufficient energy, it can heat the cell past its thermal stability limits and cause cell thermal runaway.

Internal faults related to gross manufacturing defects usually occur very early in the life of a cell. These types of failures can occur on manufacturer assembly lines where cells are being charged, or in the hands of consumers: a user purchases a device, plugs it into charge, and during that first charge, the cells undergo thermal runaway. These failures inevitably occur during, or immediately after, charging. There are some potential reasons for this phenomenon:

- Often early cell cycling causes dimensional changes in cell components (e.g., volume expansion) and increased pressures within a cell case. If a sharp contaminant or burr is present within a cell, dimensional changes or pressure increases may cause it to puncture separator layers and cause direct shorting.
- Charging provides electrical energy to the cell raising its state of charge, and susceptibility to thermal runaway.
- Charging provides energy to any shorting point within a cell. If a shorting point was present in a cell prior to charging, it may have caused the cell to self-discharge before sufficient heat was generated to induce thermal runaway. However, when attached to a charger, a short may draw energy continuously until thermal runaway is initiated.

There are a number of manufacturing quality control techniques that are commonly employed to detect gross defects. However, very subtle defects can escape notice during manufacturing and allow years of seemingly normal cell cycling

---

(Footnote 28 continued)

Horn QC, White KC, "Characterizing performance and determining reliability for batteries in medical device applications," ASM Materials and Processes for Medical Devices, Minneapolis, MN, August 13, 2009.

Horn QC, White KC, "Advances in characterization techniques for understanding degradation and failure modes in lithium-ion cells: Imaging of internal microshorts," Invited presentation, International Meeting on Lithium Batteries 14, Tianjin, China, June 27, 2008.

Hayes T, Horn QC, "Methodologies of identifying root cause of failures in lithium-ion battery packs," Invited presentation, 24th International Battery Seminar and Exhibit, Ft. Lauderdale, FL, March 2007.

Loud JD, Hu X, "Failure analysis methodology for Lithium-ion incidents," Proceedings, 33rd International Symposium for Testing and Failure Analysis, pp. 242–251, San Jose, CA, November 6–7, 2007.

Mikolajczak CJ, Hayes T, Megerle MV, Wu M, "A scientific methodology for investigation of a lithium-ion battery failure," IEEE Portable 2007 International Conference on Portable Information Devices, IEEE No. 1-4244-1039-8/07, Orlando, FL, March 2007.

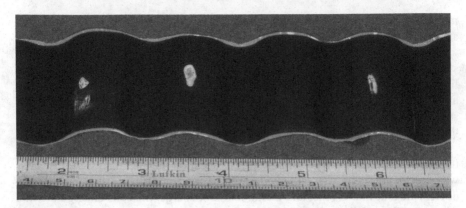

Fig. 4.11 Regions where lithium has plated on an anode are visible as white spots: upon exposure to moisture, very small, and thin deposits react to form lithium-hydroxide (a *white crystal*)

before a thermal runaway reaction occurs. Exponent has postulated[29] that any number of minor defects on a cell anode can cause very localized lithium plating. A few examples include a scratch in the anode, a point of anode delamination, a point of anode over-compression, a thin spot in the anode, or a point where a metallic contaminant has plated. Such lithium plating results in the formation of lithium dendrites and a mat of "dead lithium" composed of detached lithium dendrites. Individual dendrite shorting is usually not significant, as resistive heating quickly breaks the dendrite. However, if such shorting and heating occurs in the midst of a mat of dead lithium, it may be possible to ignite sufficient material to initiate an internal short of sufficient size to cause cell thermal runaway. Note that the amount of plated lithium postulated is in the microgram regime, and will have negligible effect on the behavior of the cell once thermal runaway is initiated (Fig. 4.11).

One important aspect of this lithium plating failure mechanism is that lithium dendrite growth only occurs during cell charging, since charging is accomplished by moving lithium ions (Li^+) and electrons from the cathode to the anode. Normally, the lithium ions intercalate into the anode safely. However, lithium ions can form lithium metal on the anode surface during charging if they are unable to

---

[29] Mikolajczak C, Harmon J, Gopalakrishnan P, Godithi R, Hayes T, Wu M, "From lithium plating to cell thermal runaway: A combustion perspective," Proceedings, 27th International Battery Seminar & Exhibit for Primary & Secondary Batteries, Small Fuel Cells, and Other Technologies, Ft. Lauderdale, FL, March 15–18, 2010.

Mikolajczak C, Harmon J, Gopalakrishnan P, Godithi R, Wu M, "From lithium plating to lithium-ion cell thermal runaway," NASA Aerospace Workshop, Huntsville, AL, November 2009.

Mikolajczak C, Stewart S, Harmon J, Horn Q, White K, Wu M, "Mechanisms of latent internal cell fault formation," Proceedings, 9th BATTERIES Exhibition and Conference, Nice, France, October 8–10, 2008.

Mikolajczak C, Harmon J, Wu M, "Lithium plating in commercial lithium-ion cells: observations and analysis of causes," Proceedings, Batteries 2009 The International Power Supply Conference and Exhibition, French Riviera, Sept 30–Oct 2, 2009.

intercalate (e.g., due to the presence of defects mentioned above). The structure of the lithium metal formation manifests itself as dendrites due to the physical nature of the process, and thus, thermal runaway failures associated with this mechanism will only occur during or immediately after charging. In Exponent's experience during investigating field failures, the majority of thermal runaway failures that occur in the field after extended "normal" use of a lithium-ion battery pack occur during, or directly after charging. The reasons for this phenomenon are:

- Lithium dendrite formation occurs during charging; thus, shorting of dendrites (which will provided highly localized heating that could potentially trigger other exothermic reactions) is most likely to occur during charging.
- Charging provides electrical energy to the cell raising its SOC, and susceptibility to thermal runaway.
- Charging provides energy to any shorting point within a cell. If a shorting point was present in a cell prior to charging, it may have caused the cell to self-discharge before sufficient heat was generated to induce thermal runaway. However, when attached to a charger, a short may draw energy continuously until thermal runaway is initiated.

Because failures due to both gross manufacturing defects and more subtle manufacturing defects generally occur during cell charging, we believe that charging cells or battery packs in bulk transport or storage should be avoided if at all possible.

Factors that Influence the Effect of Failure

The severity of a lithium-ion cell failure will be strongly affected by the total energy stored in that cell: a combination of chemical energy and electrical energy. Thus, the severity of a potential thermal runaway event can be mitigated by reducing stored chemical energy (i.e., by reducing the volume of electrolyte within a cell), or by changing the electrolyte to a noncombustible material (an area of active research but not yet commercialized). As an example, a short within a cell that has lost its electrolyte due to leakage is unlikely to result in an energetic failure. Reducing electrical energy can be done by using low capacity electrodes, or by reducing cell SOC. Finally, changing the heat transfer environment of a cell and thus affecting the removal of energy can also influence the severity of thermal runaway.

Cell Chemistry

Lithium-ion cell chemistry can affect the severity of a cell failure. Certain cathode materials allow higher energy densities than others, and cells produced from these higher energy density materials will be subject to more severe thermal runaway reactions. Cathode material reactivites are often examined and used to compare relative cathode "safety." This may be a factor in determining whether a localized fault within the cell can cause the sufficient heating to bring the entire cell to thermal runaway. However, once a cell achieves thermal runaway, the ultimate severity of the reaction is dominated by whether the cell itself will reach sufficient temperature to ignite flammable vent gases. Hot surface ignition usually requires temperatures well above gas auto-ignition temperatures—for hydrocarbons, usually in the range of 600–1,200°C (2,200°F) depending on the composition and various geometric factors of the heated surfaces.[30] Thus, unless a cathode material has sufficiently low energy density to ensure a cell remains below 600°C (1,110°F) during thermal runaway, the severity of the thermal runaway reaction will not be significantly affected by cathode chemistry.

The application of flame retardant additives to cell electrolytes and the development of non-flammable electrolytes continue to be active areas of study.[31] Although researchers have reported effective retardants or non-flammable electrolytes, these have not become commercialized due to several potential concerns such as poor cell lifetime, poor performance, and/or elevated toxicity hazards.

State of Charge

It has been observed that the vast majority of thermal runaway reactions that occur in the field occur during or shortly after cell charging. From an energy perspective, cell thermal runaway is unlikely to occur in a cell at a low SOC. Exponent's own testing showed that for many lithium-ion cells, even severe crushing of cells that are below approximately 50% SOC will not lead to a severe reaction.[32] Testing of a variety of 18650 cells at ambient temperatures has demonstrated that below 50% SOC, cell shorting will cause heating to cell case temperatures up to approximately 130°C (266°F) followed by cell cooling.

ARC testing by Exponent (see footnote 22 in chap. 1) of commercial 18650 cells of a variety of chemistries at a variety of SOC has shown that self-heating onset temperature and self-heating rate is a function of the stored electrical energy stored as chemical potential energy within the cell rather than cell chemistry

---

[30] Babrausksas V, *Ignition Handbook*, Fire Science Publishers, 2003, pp. 83–89.

[31] See for example: Dalavi S, et al., "Nonflammable Electrolytes for Lithium-Ion Batteries Containing Dimethyl Methylphosphonate," Journal of the Electrochemical Society, 157(10), A1113-A1120, 2010. Sazhin SV, Harrup MK, Gering KL, "Characterization of low-flammability electrolytes for lithium-ion batteries," Journal of Power Sources, 196(2011), 3433–3438. Feng JK, Ai XP, Cao YL, Yang HX, "Possible use of non-flammable phosphonate ethers as pure electrolyte solvent for lithium batteries," Journal of Power Sources, 177 (2008), 194–198.

[32] General approach described in IEEE 17th Annual Battery Conference on Applications and Advances paper, Loud J, Nilsson S, Du Y, "On the Testing Method of Simulating a Cell Internal Short Circuit for Lithium Ion Batteries," Long Beach, CA, 2002.

(stored electrical energy was held constant across cells of equivalent volume but varying chemistry). ARC testing by researchers at Sandia[33] showed that self-heating onset temperature can be strongly impacted by SOC. In testing one commercial cobalt-oxide cell model, the researchers found self-heating onset occurred at 80°C (176°F) for cells at 100% SOC, and at 130°C (266°F) for cells at 0% SOC. In testing of multiple cell models, Sandia researchers found that thermal runaway onset temperature is reduced for cells at increased SOC.[34] Similarly, direct electrode shorting tests (see footnote 25) have shown that reducing SOC significantly reduces the maximum temperature achieved at the point of shorting. Fire calorimetry[35] measurements (per ASTM E2058) by Ineris have shown that decreased SOC corresponds with lower peak heat-release rates for pouch cells.

Heat Transfer Environment

Finally, the heat transfer environment of a cell undergoing a thermal runaway reaction can play a large role in the severity of the reaction. High ambient temperatures or adiabatic insulation will increase the likelihood that any given internal fault can drive a cell to thermal runaway, and increase the energy available to heat the cell. Conversely, if a cell is surrounded by thermally conducting media (e.g., surrounded by densely packed cells or coolant), heat loss may prevent or mitigate a thermal runaway reaction.

If cells are assembled in close proximity, and not sufficiently heat sunk, thermal runaway in one cell can propagate to nearby cells. Exponent used this technique to initiate thermal runaway reactions (see footnote 11) in tests examining thermal runaway propagation and the effect of SOC on propagation. One cell in a battery pack at low SOC was disconnected from the other cells, charged to 100% SOC and replaced into the pack. A small heater was attached to the 100% SOC cell, and then heated locally above the melting point of its separator (approximately 150°C/300°F). At 100% SOC, ignition of cell vent gases after thermal runaway of a single cell is common, as cell case temperatures will exceed vent gas auto-ignition temperatures. Vent gas ignition in combination with the thermal runaway reaction results in measured cell surface temperatures of approximately 650°C (1,200°F). This initiating cell will then propagate thermal runaway to other cells in the battery pack, and the effect of other factors such as local heat transfer conditions can be observed.

---

[33] Roth EP, "Thermal Stability of Electrodes in Lithium-Ion Cells," Sandia Report: SAND2000-0345 J. Roth EP, Crafts CC, Doughty DH, "Thermal Abuse Studies on Lithium Ion Rechargeable Batteries," Sandia Report: SAND2000-2711C.

[34] Roth EP, "Final Report to NASA JSC: Thermal Abuse Performance of MOLI, Panasonic, and Sanyo 18650 Li Ion Cells," Sandia Report: SAND2004-6721, March 2005.

[35] Ribiere P, Laruelle S, Morcrette M, Grugeon S, Tarascon JM, Marlair G, Bertrand JP, Paillart A, "Li-ion battery: safety tests," Poster, Advanced Automotive Battery Conference (AABC).

A number of researchers have experimented with embedding cells in materials that can enhance heat transfer away from cells. Kizilel et al.[36] report thermal modeling results that suggest a phase change material produced by All Cell Technologies[37] can absorb sufficient heat from embedded cells to prevent thermal runaway propagation. NASA conducted abuse testing on battery packs incorporating a heat absorbing material placed around cells (see footnote 37 in Chap. 1). The Tesla Roadster batteries are designed with a liquid cooling system to maintain cells at uniform temperatures during normal operation and to "guarantee safety."[38]

Note that heat transfer internal to cells themselves may be slow due to the thermally insulating properties of many cell components. Thus, localized heating within large cells can be problematic, particularly during high rate discharge processes. Many manufacturers limit cell dimensions to ensure that an external short circuit will not cause sufficient internal heating to drive a cell into thermal runaway.

---

[36] Kizilel R, Sabbah R, Selman R, Al-Hallaj S, "An alternative cooling system to enhance the saftety of Li-ion battery packs," Journal of Power Sources, **194** (2009), pp. 1105–1112.

[37] http://www.allcelltech.com

[38] http://www.teslamotors.com/roadster/technology/battery

Chapter 5
Life Cycles of Lithium-Ion Cells

The typical life cycle of a lithium-ion cell is composed of approximately ten parts:

1. A cell is manufactured and undergoes initial cycling (i.e., formation) at the manufacturing facility.
2. The cell manufacturer ships the cell to a battery pack assembler or manufacturer.
3. The battery pack assembler or manufacturer combines one or more cells, protection electronics, and case materials to create a battery pack. Cell or pack testing may occur at this facility.
4. The pack assembler or manufacturer ships the battery pack to a portable electronics equipment or electric vehicle manufacturer.
5. The equipment or vehicle manufacturer installs the battery pack. Pack testing may occur at this facility.
6. The equipment or vehicle manufacture ships the device containing the battery pack to a distribution center.
7. A distribution center sells, and potentially ships the device containing a battery pack to a customer.
8. The customer uses the device with its battery pack, or re-ships the device (e.g., as a gift, as a customer return, as a mail or internet order from a retailer to a consumer, or for servicing/repair).
9. At the device end of life or battery pack end of life, the device or battery pack is discarded.[1]
10. The battery pack is transported to a solid waste disposal site or to a recycling site.

There are specific hazards associated with each of these steps.

As part of manufacturing, lithium-ion cells undergo initial cycling and aging a part of a "formation" process (discussed above). Normal formation will produce

---

[1] There are a few small volume, unique, "end-of-life" scenarios, such as satellite retirement, or disposition of human remains with embedded lithium-ion cells.

C. Mikolajczak et al., *Lithium-Ion Batteries Hazard and Use Assessment*,
SpringerBriefs in Fire, DOI: 10.1007/978-1-4614-3486-3_5,
© Fire Protection Research Foundation 2011

flammable gases which, depending upon the cell design, may or may not be vented during this process. Formation gases are vented from large format cells, prismatic cells, and pouch cells. They are not vented from small cylindrical cells such as 18650 s but rather remain contained within the cell case. In experiments concerning gas generation during the first charge of lithium-ion cells, Jehoulet et al.,[2] of SAFT detected the formation of ethylene and propylene gas, as well as small quantities of hydrogen, oxygen, nitrogen, carbon monoxide, methane, and carbon dioxide.

Should a cell contain a gross manufacturing defect that was not detected prior to initial cycling, there is a high probability that the defect will manifest itself during initial cycling as a typical infant mortality failure. Most of these failures are minor; manufacturers typically reject cells that exhibit very low capacities and very high self-discharge rates after initial cycling and aging. However, energetic failures do occasionally occur during formation. Typically, formation facilities integrate fire suppression, and thus, fires are minor and are likely to go unreported. There have been a few instances of large fires that have initiated in formation or cycling facilities that have been reported:

- November 4, 1995[3]: An explosion occurred at a Sony battery factory in Koriyama City, Japan, where cylindrical lithium-ion batteries for notebook PCs were manufactured. The fire occurred on the floor where batteries underwent final testing. Cells in this location were stored in racks 4-high under ambient temperature conditions.[4] Ultimately, approximately 3 million cells burned, 7,000 m^2 of facility was damaged and two people were injured.
- August 1997[5]: An explosion occurred at the Matsushita Battery Industry factory in Moriguchi, Osaka. The owner of the factory, T&T Dream, was a subcontractor for Matsushita. The factory carried out charge/discharge and check processes of cylindrical lithium-ion batteries. Cells in this location were stored on thirteen layers under ambient temperature conditions (see footnote 4). Ultimately, approximately 1.22 million cells burned, 1,700 m^2 of facility was burned, buildings within a 175 m radius were damaged, and two people were injured.
- August 2008[6]: A fire occurred at Batterie-Montage-Zentrum (BMZ) in Karl-stein, Germany. The fire destroyed a production area and a warehouse.

----

[2] Jehoulet C, Biensan P, Bodet JM, Broussely M, Moteau C, Tessier-Lescourret C, "Influence of the solvent composition on the passivation mechanism of the carbon electrode in lithium-ion prismatic cells," Proceedings, Symposium on Batteries for Portable Applications and Electric Vehicles, 1997.

[3] Lange L, "Squeeze on Li-ion batteries," Electronic Engineering Times, **875**, November 20, 1995, p. 1 http://findarticles.com/p/articles/mi_m0EKF/is_n2091_v41/ai_17810000/

[4] Additional information provided by the National Research Institute of Fire and Disaster (NRIFD), Japan.

[5] Hara Y, "Matsushita expects no shortage of Li-ion cells—Fire raises battery fears," Electronic Engineering Times, September 1, 1997, p. 28.

[6] Hammerschmidt C, "Fire causes heavy damage in battery factory," EE Times, August 22, 2008, http://www.eetimes.com/electronics-news/4192993/Fire-causes-heavy-damage-in-battery-factory.

- September 2008[7]: A large format lithium-ion battery that was undergoing testing at Yardney Technical Products in Pawcatuck Connecticut caught fire.

At the end of initial cell cycling and aging, cell manufacturers typically bring cells to a low to moderate SOC. This is done because manufacturers anticipate their cells may undergo extended transport and storage times prior to delivery to a customer. Properly designed and manufactured lithium-ion cells have very low self-discharge rates; commonly quoted in the range of 1–5% per month. When stored at 25°C (77°F) or below, and initially at approximately 50% SOC, a high quality lithium-ion cell can be expected to experience minimal internal impedance growth, and remain within an acceptable voltage range for many years. Storage (calendar life aging) at elevated temperatures and high voltages (high SOC) results in enhanced degradation of cell components resulting in increased internal impedance. Storage at low voltages (low SOC), and/or low temperatures, reduces the magnitude of the calendar life aging effect, and would thus, seem to indicate that storage at low voltages is preferable for maximizing cell life. However, most lithium-ion cell designs suffer from degradation if allowed to remain in a severely over-discharged state (cell voltage approximately 1 V): corrosion of copper current collectors can occur, leading to rapid impedance growth, and sometimes resulting in cell thermal runaway upon cell recharging. Thus, putting a discharged cell (at approximately 3 V) into storage is generally discouraged as extended storage periods can result in cell over-discharge. Based on these factors, lithium-ion cell manufacturers have determined that delivering cells at approximately 50% SOC is optimal for maximizing cell performance upon receipt by the customer; the reduced cell voltage reduces the effects of calendar aging, while the remaining capacity in the cell will prevent cell over-discharge for significant periods. Additionally, consumers may find convenience and satisfaction when receiving a device that has some functionality right out of the box without the need to fully charge it before its first use.

After cycling and aging, cells are packaged for transport to battery pack assembly facilities. The style of packaging must conform to packing regulations, such as those listed in the International Civil Aviation Organization Technical Instructions on the Safe Transport of Dangerous Goods by Air (ICAO Technical Instructions), and the current UN Recommendations on the Transport of Dangerous Goods (UN Recommendations). Packaging style will depend upon the type of cell being transported. Hard case cells can be bulk packaged in a densely packed configuration. Pouch cells are generally placed into individual pockets in molded trays.

Once cells reach a battery pack assembler they may be tested prior to assembly into battery packs. Such testing could be a mere measurement of open circuit voltage to detect and reject cells with high self-discharge rates (an indication of a manufacturing anomaly), or it might include cycling cells to measure capacity and

---

[7] http://www2.fluoridealert.org/Pollution/Miscellaneous/Yardney-Fire-Evacuates-Pawcatuck-Area

internal impedance. Once cells are assembled into battery packs they may undergo additional cycling, which may be used to test the pack for proper operation and to properly initialize battery pack fuel gauging devices. After production, battery packs are packaged for transport to equipment manufacturers for integration into final products.

Original equipment manufacturers (OEMs) generally install battery packs into their equipment or package them with their equipment. For complex, high value products, such as notebook computers, equipment manufacturers will generally test each device with its battery pack installed to ensure that charge and discharge within their devices functions properly (and that the battery pack is also functional). Exponent is aware of occasional fires that have occurred during final product "burn-in" testing. The battery industry encourages OEMs to ship their products with cells at 50% SOC or below; however, marketing concerns drive some OEMs to rather ship products with cells at full charge (100% SOC) so that the product is "ready to go right out of the box."

OEMs ship final products to distribution centers or retail outlets. Distribution centers or retail outlets deliver (possibly through air shipments) products to individual customers. Customers may reship products for any number of reasons, including repair. Units returned for repair could be at any SOC. Best Buy (see footnote 2 in Chap. 2) reports that "… a consumer, before concluding that a product requires service, often will have endeavored to use it under fully charged conditions."

Safety during transport, particularly by air, received significant attention in 2010 due to a proposed rulemaking by the US Pipeline and Hazardous Materials Safety Administration (PHMSA), Docket No. PHMSA-2009-0095 (HM-224F). As a result, numerous cell manufacturers, OEMs, and shipping companies submitted comments to PHMSA regarding transport of lithium-ion batteries.[8] Comments by entities such as FedEx, UPS, Motorola, Best Buy, and Panasonic, who have hundreds of years of combined shipping experience, and who have shipped billions of lithium-ion batteries,[9] provide an overview of transport practices, transport volumes, and transport safety.

Transport Practices

Large shipments of lithium-ion cells, battery packs, or battery packs contained in equipment are accomplished on pallets, and occasionally in unit load device (ULD) containers. Pallets typically contain multiple layers of boxes and may be enclosed in a cardboard over-pack, wrapped in plastic, or netted to secure the

---

[8] PHMSA-2009-0095.

[9] IATA estimates the number of Li-ion and lithium primary cells shipped by air in 2008 was approximately 1.2 billion. PHMSA-2009-0095-0047.1, The Association of Hazmat Shippers estimates that 3.3 billion cells were shipped in 2008: PHMSA-2009-0095-0050.1.

boxes to the pallet. Smaller shipments of individual boxes (US domestic) are sent through shipping companies such as UPS and FedEx.

Many shipments are by air for a number of reasons:

- The majority of cells or battery packs are produced in Asia.
- Fast consumer electronics design cycles and "just-in-time" practices (such as those used in the medical device industry) necessitate rapid transport of product.
- Consumers often request next-day or 2-day shipping of devices ordered online.
- Remote US locations such as Alaska, may not be able to tolerate the long waits for ocean-based shipments, and at times can be minimally accessible or inaccessible by roads due to weather conditions.

Transport Volumes

A number of estimates have been made regarding lithium-ion transport volumes. Some estimates are listed below.

- PRBA reports[10]:
  - "... in 2008 over 3 billion lithium ion cells were manufactured worldwide"
  - "In 2009 there were nearly 340 million notebooks, cellular phones, and digital still and video cameras packed with or containing lithium-ion batteries that were shipped to the U.S."
- The International Air Transport Association (IATA) estimates (see footnote 9) the number of lithium-ion and lithium primary cells shipped by air in 2008 was approximately 1.2 billion.
- The Association of Hazmat Shippers estimates (see footnote 9) that 3.3 billion cells were shipped in 2008.
- UPS reports just seven of their customers are responsible for over 40-million lithium battery containing packages per year, including:
  - A single medical device shipper that sends over 750,000 annual shipments of life-sustaining devices.
  - A single camera equipment shipper that sends over 7.9 million annual shipments.
  - Three cell phone companies that ship a combined 16-million packages by air a year.
  - A single laptop supplier that transports 14.6 million laptops by air per year.

---

[10] PHMSA-2009-0095-0117.

UPS uses driver tools called delivery information acquisition devices (DIAD) that contain lithium-ion batteries. These units are periodically shipped for repair or service. UPS uses 130,000 DIAD worldwide. UPS services more than 14,000 units per month at three repair sites and many of these units are shipped by air.

The National Funeral Directors Association (NFDA)[11] estimates approximately 1% of human remains contain an implanted medical device with a lithium or lithium-ion battery. This accounts for about 1,000 human remains with such a device being transported by air each year.

Transport Safety

There have been reports of lithium-ion battery fires during transport. The FAA has assembled a list of fires associated with aircraft transport.[12,13,14] Similar lists have been published by the UN Subcommittee of Experts on the Transport of Dangerous Goods,[15] and by the ICAO Dangerous Goods Panel.[16] Descriptions of the aircraft transport incidents are provided in two tables: Table 5.1 lists failures that occurred to lithium-ion batteries during transport as air cargo, while Table 5.2 lists failures that occurred to personal lithium-ion batteries during transport.

One common theme to the air cargo incidents is improper packaging of the lithium-ion batteries involved in the incident. There is considerable consensus in the industry that shipping is safe when shippers comply with existing regulations for lithium-ion battery shipping—specifically those regulations derived from the UN Recommendations to package devices to prevent short-circuiting. The Air Transport Association (ATA) reports,[17] "... there has not been a single in-flight incident attributed to a commercial shipment of properly packaged electronic devices containing batteries ..." in the last 5 years. PRBA reports (see footnote 10), "There has never been a fire on an aircraft attributable to lithium ion cells, batteries, or the products into which they are incorporated where existing

---

[11] PHMSA-2009-0095-0173.

[12] FAA Office of Security and Hazardous Materials, "Batteries & Battery-Powered Devices, Aviation Incidents Involving Smoke, Fire, Extreme Heat or Explosion, incidents recorded as of March 20, 1991, through August 3, 2010, http://www.faa.gov/about/office_org/headquarters_ offices/ash/ash_programs/hazmat/aircarrier_info/media/Battery_incident_chart.pdf.

[13] Richard B, "Lithium Battery Update," Office of Hazardous Materials Safety, Pipeline and Hazardous Materials Safety Administration, US Department of Transportation, September 2009.

[14] Webster H, "Lithium Battery Update, Recent Battery Incidents," FAA, November 17, 2009.

[15] UN/SCETDG/31/INF.41, Committee of Experts on the Transportation of Dangerous Goods and on the Globally Harmonized System of Classification and Labeling of Chemicals, Sub-Committee of Experts on the Transport of Dangerous Goods, 31st Session, Geneva, 2–6 July 2007, Item 3 of the provisional agenda.

[16] DGP/22-IP/4, Dangerous Goods Panel (DGP) 22nd Meeting, Montreal, 5–16 October 2009, Enhanced Requirements for the Transport of Lithium Batteries.

[17] PHMSA-2009-0095-0077.1.

Table 5.1 Air cargo transport incidents (see footnotes 12–16)

Date	Incident description	Likely cause of failure
August 2009	FedEx discovered a burning and smoking package at one of their facilities, it contained 33 GPS tracking devices with lithium-ion batteries, two of the devices had heated causing surrounding packaging and cushioning to ignite. The package was not properly labeled.	Mechanical shock/vibration External short circuit Improper packaging
August 2009	UPS found a smoldering package at its Taiwan Hub. Inspection of other packages in the same consignment indicated that similar batteries were shipped without terminal protection.	External short circuit Mechanical shock/vibration Improper packaging
July 2009	UPS found a package emitting smoke in the Dominican Republic. The package had arrived from Romulus, MI. It contained numerous loose lithium-ion cell phone batteries, not protected from short-circuiting. The package documentation indicated, "used batteries."	External short circuit Improper packaging
June 2009	UPS found a charred and black package inside a ULD that was being unloaded in Honolulu, HI. The package had traveled from New Jersey via Philadelphia and California. The package contained an "e-bike" battery, composed of lithium-iron phosphate cells. The cardboard packaging and inner bubble wrap material was largely intact. FAA investigation determined that external short-circuiting of the battery pack caused overheating of circuitry. Cells swelled but did not vent or ignite.	External short circuit brought about by a combination of transport and handling shock and vibration with improper packaging
August 2008	UPS discovered a smoking package containing lithium-ion battery powered LED lamps at a ground sort facility.	External short circuit brought about by a combination of transport and handling shock and vibration with improper packaging
December 2007	Package containing an RC helicopter kit with lithium-ion polymer batteries was discovered emitting smoke at a FedEx sort facility.	External short circuit brought about by a combination of transport and handling shock and vibration with improper packaging

(continued)

Table 5.1 (continued)

Date	Incident description	Likely cause of failure
December 2007	A customs inspector cut into a box with a knife, and accidentally cut into a lithium-ion polymer battery pack. The package contained lithium polymer batteries for RC aircraft, and was improperly manifested/packaged.	Mechanical damage: puncture
September 2007	FedEx discovered a box emitting smoke on offload. The box contained three inner fiberboard boxes. Each inner box contained 120 lithium-ion batteries. The fire was contained to one inner box.	
August 2007	During customs inspection, one of 440 lithium-ion polymer batteries in a package began burning.	Mechanical damage
November 2006	Batteries selected for inspection by a US customs officer underwent thermal runaway. Batteries had arrived in US from China.	Mechanical damage
July 2006	Unlabeled and unmarked package was discovered by FedEx to have caught fire while being held in bond for customs clearance in Korea. The package had traveled from Vienna via Paris and Subic Bay.	External short circuit brought about by a combination of transport and handling shock and vibration with improper packaging
June 2006	Aircraft cargo hold fire alarm was activated during taxiing for departure. The captain activated fire suppression and passengers were evacuated. The source was found to be a package of lithium polymer batteries. The shipment was declared as electric parts (violation of shipping requirements). No UN test report was available for the batteries.	External short circuit brought about by a combination of transport and handling shock and vibration with improper packaging
March 2006	FedEx discovered a package releasing smoke in an outbound station in China. The package contained lithium-ion batteries.	External short circuit brought about by a combination of transport and handling shock and vibration with improper packaging
June 2005	UPS discovered a burned package in California upon unloading a ULD from Shanghai. The package contained a lithium-ion battery pack. Upon discovery, the package and contents were cool to the touch and no smoldering was evident.	External short circuit brought about by a combination of transport and handling shock and vibration with improper packaging

(continued)

Table 5.1 (continued)

Date	Incident description	Likely cause of failure
August 2004	A box containing two lithium-ion battery modules for an electric vehicle prototype suffered fire damage. The box was found when FedEx cargo handlers detected smoke emanating from a ULD on an aircraft loading ramp in Memphis, TN. The battery modules had been packaged with metal tools. Investigators[a] concluded that the tools had caused external shorting of the cells. Investigators also concluded that the batteries had not been packaged per DOT requirements in a manner to prevent short circuits.	External short circuit brought about by a combination of transport and handling shock and vibration with improper packaging

[a] National Transportation Safety Board, Hazardous Materials Accident Brief, Accident # DCA04MZ001, http://www3.ntsb.gov/publictn/2005/HZB0501.pdf

regulations were complied with." UPS reports,[18] "UPS has experienced some incidents involving packages of lithium batteries that overheated or were involved in fires, and has individually analyzed each event … UPS is aware of no instance in which the batteries responsible for these incidents were offered in compliance with the applicable regulations in effect at the time of shipment. Defects ranged from poorly designed or assembled batteries that allow short circuits or other faults, to unsafe packaging, to flaws in equipment containing lithium batteries."

Another commonality among the air cargo incidents is that packages have often been found smoldering in ULD devices on *offload* from aircraft, implying that thermal runaway of cells initiated in-flight. In many instances, the damage has remained isolated to the initiating package. The very limited propagation of these failures is consistent with shipping at low SOC, and with limited airflow to affected cells because of packaging.

The FAA data can be used to estimate a typical failure rate of lithium-ion cells in air cargo transport. Assuming that annual cell production is approximately 3-billion cells (per 2008 estimates), and that all of these cells are transported by air once (a reasonable, and likely conservative assumption since many cells are subjected to multiple air shipments), then failures that produce smoke and heating occur at a rate of approximately 1 in 1-billion cells shipped. In contrast, field failures of cells (battery packs personal use) tend to occur more frequently, and to occur when the packs are fully charged. Typically, in the US, failure rates worse than 1 in 1-million trigger recall actions, as it is generally assumed that normal cells from reputable manufacturers will exhibit failure rates better than 1 in 1

---

[18] PHMSA-2009-0095-0062.

Table 5.2 Personal battery pack air transport incidents (see footnotes 12–16)

Date	Description	Likely cause of failure
April 2010	A lithium-ion battery powered curling iron in checked baggage appears to have activated and caused thermal runaway of a spare lithium-ion battery. The bag and some contents were scorched.	Unintended device activation, followed by external heating of cells
September 2009	An air carrier's battery available for on-board use by passengers was dropped in-flight and caught fire. Flight attendants attempted to extinguish fire with a Halon extinguisher, and then by pouring water over the battery pack.	Mechanical shock
August 2008	Passenger found his notebook computer battery smoking—he gave it to a flight attendant who placed it in a coffee pot in the galley and poured "water and Sprite" on it.	Cell internal fault
March 2008	i-Theater Video Display unit containing a lithium-ion polymer battery pack underwent thermal runaway in-flight. Captain doused it with water.	Cell internal fault
June 2007	While waiting at a gate area, a passenger plugged his notebook computer into an electrical outlet. The computer began smoking and eventually burst into flames.[a] Fire extinguishers were used to suppress the fire.	Cell internal fault
May 2007	Ramp worker removed a checked bag that was on fire when loading a passenger aircraft. Fire department investigation indicated source of fire was a battery pack for a handheld video game.	Cell internal fault or mechanical damage
September 2006	Prior to departure, a passenger's notebook computer began to smoke. It was removed from the airplane to the gate area where it continued to smoke and a small flame appeared; a customer service representative discharged a fire extinguisher on the fire.	Cell internal fault
May 2006	Spare notebook computer battery pack purchased on eBay, and placed in hand luggage in an overhead bin underwent thermal runaway. Incident occurred before flight departure, and crew used extinguishers on the plane before the battery pack was removed from aircraft. The fire was eventually suppressed by the fire department after reigniting once.	Cell internal fault or mechanical damage

[a] http://www.youtube.com/watch?v=mlZggVrF9VI

million.[19] This difference in failure rate relates to a difference in the failure mechanisms that can affect new cells under shipping (and storage) conditions compared with cells that are in use. Examination of available incident data suggests that air cargo incidents are caused by mechanical damage or external short circuiting of cells, while incidents with personal battery packs are often caused by cell internal faults.

Storage Facility Safety

Lithium ion cells, battery packs, and equipment containing lithium-ion battery packs will likely be stored in warehouses in many stages of production and distribution. For example, cells may be stored at cell manufacturer warehouses, at distributor warehouses, at pack assembler warehouses, and at various intermediate locations such as customs warehouses. Packs and equipment containing lithium-ion batteries may also be stored at retailer locations. Some of the hazards associated with warehouses are similar to those encountered during transport. There is a potential for mechanical damage due to poor handling such as boxes or pallets being dropped or damaged by forklift accidents. Crush or puncture damage to cells or battery packs can lead to release of electrolyte, short circuiting, and possibly cell thermal runaway that can result in a fire. There is also potential for external heating of the cells due to a fire initially unrelated to lithium-ion battery packs that ultimately results in venting or thermal runaway of the cells. Storage at reduced SOC reduces the likelihood that crush, puncture, or external heating will lead to cell thermal runaway and a fire ignited by heated cell cases. Nonetheless, if electrolyte is released, or cells vent, the released gases pose potential toxicity and flammability/explosion hazards.

If cells are not being charged while in storage, or otherwise being handled, the likelihood of spontaneous cell thermal runaway occurring is very low, particularly if cells are stored at reduced SOC. However, battery charging is likely to occur at some storage facilities, particularly at facilities where discharged battery packs are

---

[19] Exact failure rates of lithium-ion cells and battery packs in the field are not published. Due to confidentiality requirements, the CPSC publishes very limited information regarding the circumstances of failures that have triggered lithium-ion cell recall actions. Failure rates are not published, nor are specific details regarding individual battery failures. Occasionally battery failures are reported in news stories. However, these reports generally do not contain details sufficient to make a determination regarding the cause of failure, or the rate of failure or the products described. In addition, some incidents reported in the news have later been found to have been the result of user abuse rather than a defect in the battery. OEM investigations of failed product are generally kept confidential. Therefore, it is very difficult to determine the rate of failure of cells in the field. However, in Exponent's experience, when OEMs work with the CPSC to determine if a recall on a lithium-ion battery is warranted, barring evidence of a specific manufacturing defect, if failure rates are below 1 in 1-million cells, CPSC generally agrees that a recall is not warranted.

charged in preparation for installation into vehicles, for example, at service stations, or at battery pack switching stations such as those being demonstrated and constructed for Better Place vehicles (see footnote 18 in Chap. 2). As discussed previously, although cell thermal runaway reactions are rare, in the field, they are most likely to occur during charging or immediately after battery pack charging.

Recycling

At end of life, users dispose of lithium-ion batteries into municipal waste streams, donate used equipment such as cell phones to charities, or attempt to recycle used batteries. In the US, nickel cadmium (NiCad) and lead acid batteries have long been classified as universal waste, and recycling of these batteries has been required. In order to facilitate NiCad recycling, the battery industry created the Rechargeable Battery Recycling Corporation (RBRC).[20] RBRC provides collection of rechargeable batteries at a wide range of retail locations (ideally, locations where consumers may have purchased rechargeable batteries). Rechargeable batteries in a variety of chemistries are accepted, and lithium-ion cells are becoming a greater part of the mix of batteries being processed. RBRC estimates[21] collection of 1.5-million pounds of lithium-ion batteries in 2009. This volume is expected to increase as more states pass mandatory recycling laws.[22]

Under the RBRC Program, battery and cell phone collection boxes are located in stores such as: RadioShack, Target, Sears, K-Mart, RiteAid, Walgreens, Home Depot, Lowes, Orchard Supply Hardware, Verizon Wireless, and many others. RBRC operates (see footnote 23) approximately 30,000 collection locations in the US and Canada. There is no charge for drop-off of small quantities of batteries or cell phones. The batteries are placed into individual sealed bags to prevent short-circuiting, and placed into a collection box. Boxes are transported via UPS to an RBRC sorting center, where batteries are sorted by chemistry. From the sorting center, batteries are transported to recycling facilities.

Lithium-ion batteries submitted for recycling may contain appreciable electrical energy, as well as chemical energy. Discarded batteries from single cell applications will likely have greatly reduced capacities. However, since performance of multi-cell battery packs is determined by the capacity of the weakest cell in a pack, a discarded multi-cell pack many contain a number of cells that have retained considerable electrical energy. Thus, protecting cells from short-circuiting and severe mechanical damage during transport to recycling facilities and handling

---

[20] http://www.call2recycle.org/home.php?c=1&w=1&r=Y

[21] PHMSA-2009-0095-0181.

[22] At the time of this writing, both California and New York have regulations that ban or will ban disposal of lithium-ion batteries in municipal waste. See Kerchner G, "Regulatory and Legislative Update," Proceedings, 28th International Battery Seminar and Exhibit, Ft. Lauderdale, FL, March 14–17, 2011.

Fig. 5.1 Two examples of typical battery recycling/collection bins

while at those facilities remains important. At the time of this writing 49 CFR 173.185 specifies that cells and batteries being transported for recycling are Class 9 Hazardous Material. They are exempted from UN testing requirements and requirements for UN specification packaging if protected against short circuits.

Until recently, there has been very little regulatory guidance regarding shipping batteries for disposal. It has been common practice for batteries of all chemistries to be simply dropped into collection bins (Fig. 5.1). Most batteries collected this way are spent non-rechargeable batteries with limited remaining capacity. However, fires have ignited when large bins full of loose batteries (without taped electrodes) have been in transit to recycling centers (see footnote 13). There have also been reports of fires at recycling centers where large volumes of battery packs must be sorted and transported throughout the facilities.[23,24] In April 2009, PHMSA issued a battery enforcement letter discussing the hazards associated with battery recycling. The battery and recycling industries are attempting to address these issues as well as the issue of transporting damaged batteries by developing harmonized international dangerous goods regulations for waste and damaged lithium batteries. Industry groups have submitted a draft proposal to the UN transport subcommittee, on which action is expected before the end of 2012. Industry groups expect new regulations to be effective in 2015 (see footnote 1 in Chap. 3).

---

[23] http://blogs.edmunds.com/greencaradvisor/2009/11/fire-explosions-destroy-canadian-lithium-battery-recycling-plant.html

[24] http://www.wsbtv.com/news/27570553/detail.html   and   http://wbhf.wordpress.com/2011/04/18/packing-of-batteries-could-be-to-blame-for-plant-fire/

Chapter 6
Lithium-Ion Fire Hazard Assessment

Although numerous studies of lithium-ion cell thermal stability exist, these studies are generally concerned with reactions that occur within a cell up to the point of thermal runaway. There has been relatively little work conducted to examine electrolyte or vent gas combustion properties (see footnote c in Table 1.1 of Chap. 1), or the fire hazards associated with thermal runaway reactions. The bulk of the publically available work in this area has been conducted by researchers at Sandia, the FAA, the Civil Aviation Authority of the United Kingdom (CAA), the Naval Research Laboratory,[1] and at Exponent. Researchers at Sandia conducted basic flammability tests on cell vent gases (spark ignition tests), and conducted chemical analysis of cell vent gases to identify their components. Researchers at the FAA, the CAA, the Naval Research Laboratory, and at Exponent[2] conducted studies to assess potential fire hazards associated with transport and storage of lithium-ion batteries, as well as the effectiveness of various suppressants. However, in all of these instances, published data has included only small-scale testing results. Testing has been conducted with single cells, relatively small quantities of cells (usually less than 100), or small battery packs: usually notebook computer battery packs. There are no large-scale testing results such as those designed to determine lithium-ion cell commodity specifications or sprinkler system (water based or other) configurations or design criteria published to date.

The small-scale testing conducted to date does provide a number of insights regarding the fire hazards associated with lithium-ion cells and battery packs, and therefore the results of these studies are discussed here. Recommendations for

---

[1] Naval Research Laboratory test capabilities include obtaining compartment temperature profiles, obscuration data, thermal radiation measurements, heat release rates, and gas samples. However, published data from Navy testing is sparse. See Williams F, Winchester C, "Lithium Battery Shipboard Safety," April 24, 2010.

[2] Mikolajczak CJ, Moore D, "A study of passenger aircraft cargo hold environments," Exponent Failure Analysis Associates, Inc., May 2001; http://www3.ntsb.gov/events/2006/PhiladelphiaPA/Exhibits/350563.pdf.

C. Mikolajczak et al., *Lithium-Ion Batteries Hazard and Use Assessment*,
SpringerBriefs in Fire, DOI: 10.1007/978-1-4614-3486-3_6,
© Fire Protection Research Foundation 2011

large-scale testing are made in Chap. 7. In assessing overall fire protection strategies, data collected from small-scale testing should be used with caution and
validated at the full scale when assessing suppression system design criteria.

Flammable Cell Components

The most flammable component of a lithium-ion cell is the hydrocarbon-based
electrolyte. The hydrocarbon-based electrolyte in lithium-ion cells means that
under fire conditions, these cells will behave in a fundamentally different way than
lead acid, NiMH or NiCAD cells, which contain a water-based electrolyte.
Although all charged cells contain stored electrical energy, even fully discharged
lithium-ion cells contain appreciable chemical energy that can be released through
combustion of the electrolyte. Water-based chemistries, under some charging
conditions can produce hydrogen gas through electrolysis of the water; however,
this hazard is seldom a concern during storage where no charging occurs. If cells
with water-based electrolyte are punctured or damaged, leakage of the electrolyte
can pose a corrosive hazard; however, it does not pose a flammability hazard. In
comparison, leakage or venting of lithium-ion cells will release flammable vapors.
If fire impinges on cells with water-based chemistries, the water in the cells
absorbs heat and reduces the total heat release of the fire. In comparison, fire
impingement on lithium-ion cells will cause release of flammable electrolyte,
increasing the total heat release of the fire (assuming well-ventilated conditions),
and possibly increase the total heat release rate of the fire.

Other combustible components in a lithium-ion cell include a polymeric
separator i various binders used in the electrodes, and the graphite of the anode.
Some of these components will degrade if a cell undergoes thermal runaway and
produce flammable gases that will vent from the cell. Lithium-ion cells do not
contain metallic lithium in any significant quantity to affect fire suppression; in
lithium-ion cells, Li^+ ions function simply as carriers of electric charge. In contrast, lithium primary (lithium metal) batteries contain a significant mass of
metallic lithium as their anode material.

In 2000, Crafts, Borek, and Mowry[3] described the composition of vent gas from
a noncommercial cell subject to heating at a rate of 1°C/min (1.8°F/min) up to a
temperature of 200°C (392°F). The tested cells had an NCO cathode, blended
graphite anode, and an electrolyte of blended EC and DEC. They found that the
vent gas included hydrogen, carbon monoxide, carbon dioxide, methane, ethylene,
ethane, propylene, and C4 and C5 hydrocarbons. A large proportion of the vent gas
was carbon dioxide. Crafts et al., did not report the relative quantities of each
compound produced during cell venting.

---

[3] Crafts C, Borek T, Mowry C, "Safety Testing of 18650-Style Lithium-ion Cells," Sandia
National Laboratories, SAND2000-1454C, May 2000.

Table 6.1 Normalized gas composition of vented cells from Sandia testing (without N_2, O_2, or Ar)

Cell type	100% SOC fresh cell ARC to 160°C vented 130°C	100% SOC fresh cell ARC to 160°C vented 130°C	100% SOC aged cell ARC to 160°C Pre punctured	60% SOC aged cell ARC to 160°C vented 130°C	60% SOC aged cell ARC to 160°C vented 130°C
Max sample temp	130°C	160°C	160°C	160°C	160°C
Gas species	Volume percent (%)				
H_2	5.1	5.9	6.5	5.0	7.3
CO	15.1	6.4	8.4	6.5	9.1
CO_2	61.4	75.8	68.0	66.0	58.4
CH_4	7.4	1.9	1.2	2.0	2.4
C_2H_4	8.7	8.8	15.5	19.0	15.7
C_2H_6	1.9	1.1	0.3	1.5	1.4
Ethyl fluoride	ND	ND	ND	ND	5.6
Propylene	0.3	0.1	0.0	0.0	0.0
Propane	0.0	0.0	0.0	0.0	0.0

ND indicates none detected

In 2004, Sandia researchers released a much more extensive study of cell vent gases.[4] The Sandia researchers conducted a study of electrolyte gas decomposition species by testing typical electrolyte components alone and in combination with ARC and TGA apparati. They sampled headspace gases from prototype cells, gases generated within cells prior to cell venting, and gases released by cells during cell venting. Test results (Table 1.2 and Table 6.1) were similar in nature to those reported in 2000. However, during this testing the Sandia researchers reported on the formation of ethyl fluoride (C_2H_5F) under certain test conditions. In this document, Sandia researchers reported vent gas volumes and relative vent gas component percentages.

When a cell vents, the released gases will mix with the surrounding atmosphere, and depending upon a number of factors including fuel concentration, oxygen concentration, and temperature, the resulting mixture may or may not be flammable. The flammability limits of a gaseous fuel/air mixture are the prime measures for ascertaining whether that mixture is flammable. Fuel/air mixtures have two flammability limits: a lower flammability limit (LFL) or lean limit, below which the concentration of fuel is too low to allow flame propagation, and an upper flammability limit (UFL) or rich limit, where the concentration of fuel is too high for the available oxygen to support flame propagation. If the fuel concentration in a particular gas mixture is between the LFL and UFL, that mixture is ignitable. If a competent ignition source is present, a flame can

[4] Roth EP, Crafts CC, Doughty DH, McBreen J, "Advanced Technology Development Program for Lithium-Ion Batteries: Thermal Abuse Performance of 18650 Li-Ion Cells," Sandia Report: SAND2004-0584, March 2004.

Table 6.2 Flammability limits of fuel/air mixtures[a]

Compound	Lower flammability limit (fuel volume %)	Upper flammability limit (fuel volume %)
Hydrogen	4.0	75.0
Carbon monoxide	12.5	74.0
Methane	5.3	15.0
Ethylene	3.1	32.0
Ethane	3.0	12.5
Propylene	2.4	10.3
C4 hydrocarbons	~1.6–1.9	~ 8.4–9.7
C5 hydrocarbons	~ 1.4–1.5	~ 7.5–8.7

[a] For atmospheric pressure, room temperature, and upward propagation in a tube, see: Lewis B, von Elbe G, *Combustion, Flames and Explosions of Gases*, 2nd Edition, Academic Press, New York, 1961

propagate through the mixture. If the fuel concentration in a particular gas mixture is outside the range bounded by the LFL and UFL, then that mixture will not be ignitable. At each limit, the scarcity of one reactant results in a rate of heat generation that is just low enough to be exactly balanced by the rate of heat transfer away from the reaction zone.

Every fuel has unique flammability limits in a specific oxidizing atmosphere, under specific conditions of temperature and pressure. These limits are determined by the fuel's specific combustion chemistry and the heat transfer properties of the surrounding atmosphere. Since the details of combustion chemistry are complex, flammability limits are determined empirically with standardized tests.[5] Table 6.2 lists flammability limits for various fuel/air mixtures of components found in lithium-ion cell vent gases at atmospheric pressure and room temperature. Although testing has shown that lithium-ion cell electrolyte mixtures and vent gases are ignitable, specific flammability limits of these mixtures have not been determined.

Oxygen concentration, inert diluent composition, temperature, pressure, and the presence of specific suppressant chemicals affect flammability limits. If oxygen concentration of the mixture drops due to its replacement by a specific inert diluent, such as nitrogen, carbon dioxide, or non-combustible products of combustion, the flammability limits of the mixture narrow until the oxygen concentration drops to a level below which a flame will not propagate, regardless of the fuel concentration. The flammability limits of carbon monoxide, methane, ethylene, and propylene narrow as oxygen concentration is reduced by the addition of excess

---

[5] ASTM E681 describes a standard test method for determining flammability limits.

Table 6.3 Maximum safe percentage of oxygen in mixtures of combustibles with air and carbon dioxide or nitrogen[a]

Compound	Volume % of oxygen with carbon dioxide diluent	Volume % of oxygen with nitrogen diluent
Hydrogen	5.9	5.0
Carbon monoxide	5.9	5.6
Methane	14.6	12.1
Ethylene	11.7	10.0
Ethane	13.4	11.0
Propylene	14.1	11.5
C4 and C5 hydrocarbons	14.5	12.1

[a] For atmospheric pressure, room temperature, and upward propagation in a tube, see: Lewis B, von Elbe G, *Combustion, Flames and Explosions of Gases*, 2nd Edition, Academic Press, New York, 1961

inert gases such as carbon dioxide; which can represent a large fraction of vent gas composition.[6]

The maximum oxygen concentration at which the mixture will not be flammable at any fuel concentration is referred to as the "maximum safe percentage of oxygen." In general, increasing the initial gas temperature of a fuel/air mixture results in reduced heat losses from reactions; thus, the flammability limits of that gas mixture broaden. Lowering atmospheric pressure has a minimal effect on the flammability limits of fuel/air mixtures until a very low pressure has been achieved. Until the gas pressure is reduced below 3 psia (approximately 11.7 psi on a pressure gauge at sea level), the flammability limits are only slightly affected, although the total heat release will be reduced proportionately with the air pressure.

Table 6.3 lists maximum safe percentages of oxygen in mixtures of combustibles with air and carbon dioxide or nitrogen at atmospheric pressure and room temperature. (With no added diluent, air contains approximately 21% oxygen.)

In general, increasing the initial gas temperature of a fuel/air mixture results in reduced heat losses from reactions; thus, the flammability limits of that gas mixture broaden.[7] Lowering atmospheric pressure has a minimal effect on the flammability limits of fuel/air mixtures until a very low pressure has been

---

[6] Data on the effect of dilution with carbon dioxide on carbon monoxide and methane flammability limits can be found in Lewis B, von Elbe G, *Combustion, Flames and Explosions of Gases*, 2nd Edition, Academic Press, New York, 1961. Data on the effect of dilution with carbon dioxide on ethylene and propylene flammability limits can be found in Coward HF, Jones W, *Limits of Flammability of Gases and Vapors*, Bulletin 503, US Bureau of Mines, 1952.

[7] Data on the effect of initial gas temperature on the flammability limits of methane can be found in Wierzba I, Ale BB, "The Effect of Time of Exposure to Elevated Temperatures on the Flammability Limits of Some Common Gaseous Fuels in Air," Journal of Eng. for Gas Turbines and Power, **121**, January 1999, pp. 74–79.

achieved.[8] Until the gas pressure is reduced below 3 psia (approximately 11.7 psi on a pressure gauge at sea level), the flammability limits are only slightly affected, although the total heat release will be reduced proportionately with the air pressure.

Finally, the presence of halogenated compounds in small quantities can significantly narrow flammability limits (as discussed below with regard to Halon compounds). Thus, the presence of halogenated hydrocarbons such as ethylene fluoride could significantly affect vent gas flammability.

Determination of specific flammability limits of vented cell electrolyte remains a gap in knowledge and is discussed further in Chap. 7.

Stored Energy (Chemical and Electrical)

The energy content of a lithium-ion cell will be a combination of the stored electrical energy and the stored chemical energy. Stored electrical energy is straightforward to measure, and should be similar to the capacity rating of the cell or battery pack, usually expressed in Ah. Using rated cell capacity is a good approximation, although it will not provide an exact measure of stored electrical energy for a few reasons.

- Nominal capacity ratings are not exact, but are often set based on expected aging behavior. Typically, a cell is required to maintain 80% of nominal capacity after 300 cycles. Thus, initial cell capacity may exceed nominal cell capacity in order to meet cell-aging specifications.
- Cell capacity ratings usually assume that discharge of individual cells is terminated at approximately 3 V rather than 0 V. However, for traditional lithium-ion cells very little energy is stored in the voltage range between 0 and 3 V.
- Cell capacity fade occurs with usage and storage. Thus, the capacity of a used cell will likely be below nominal capacity.

Cell state of charge can easily be accounted for in calculations of expected energy release.

Cell stored chemical energy is not straightforward to measure. The most common approach to measurement of total heat release from a device or material for fire protection purposes[9] involves oxygen consumption calorimetry, often associated with a cone calorimeter. Oxygen consumption calorimetery is based on the assumption that reduced oxygen concentrations measured in the exhaust duct of the calorimeter are the result of oxygen consumption due to combustion

---

[8] Data on the effect of atmospheric pressure on the flammability limits of natural gas/air mixtures can be found in Lewis B, von Elbe G, *Combustion, Flames and Explosions of Gases*, 2nd Edition, Academic Press, New York, 1961.

[9] Babrauskas V, Grayson SJ (eds), *Heat Release in Fires*, E&FN Spon, New York, 1992.

processes, and that the energy released by complete combustion per unit mass of oxygen consumed is a constant. These assumptions have been shown to be reasonable for most common combustibles burning in air at ambient pressures. However, lithium-ion cell vent gases contain significant percentages of CO_2 formed from thermal degradation processes (pyrolysis rather than combustion). This form of CO_2 production (breaking of a C–C bond in a carbonate compound to release an O–C–O functional group) does not involve the same energy release as the formation of CO_2 from typical combustion processes (reaction of CO and OH). Therefore, oxygen consumption calorimetry is likely to significantly over-predict heat release from lithium-ion cell venting and combustion and is not appropriate, without significant modification, for determining heat release rates of lithium-ion cells.[10] Heat release rate measurements represent a gap in understanding of the fire hazard and are discussed further in Chap. 7.

An approximation of stored chemical energy can be made by considering the heats of combustion of various flammable components of the cell. Exponent estimates that on average an 18650 cell will contain 3–6 g of electrolyte but could contain up to 10 g. To be conservative, we can assume[11] that an 18650 cell contains approximately 10 g of electrolyte and roughly 1.6 g of separator (and other insulators), which is usually made from polyethylene, polypropylene, or a combination of these two materials. Since these materials are both plastics with comparable heats of combustion, we can treat the separator as being composed of polypropylene (approximately 42.66 kJ/g). Commercial lithium-ion cell electrolytes are composed of a mixture of organic carbonate solvents such as diethyl carbonate (DEC). Since the heat of combustion of DEC (20.92 kJ/g) is comparable to other organic carbonate solvents, the heat of combustion of DEC can be used to estimate the chemical energy of the electrolyte. From this calculation, the expected heat of combustion (heat output from the combustion) of an individual 18650 cell is approximately 280 kJ. The individual cells in the battery pack likely contain between 7 and 11 Wh (25 and 40 kJ) of energy when fully charged. Thus, an estimate of the total energy (electrical plus chemical) that can be released by one fully charged 2- to 3-Ah 18650 cell is approximately 300–320 kJ, while an 18650 cell at approximately 50% SOC would contain a total energy between 290 and 300 kJ.

Using an estimate of approximately 1–5 g of electrolyte and ½ to 1 g of separator per Ah of cobalt-oxide cell capacity (or approximately 100–150 kJ per Ah of cell capacity) is likely appropriate for small commercial cobalt-oxide based lithium-ion cells of various form factors, as the internal construction of these cells tends to be similar: electrode and separator thicknesses are similar, electrolyte is designed to be largely absorbed into the electrode materials with little free

---

[10] The experimental result can be compared to the spurious result that would be produced if a CO_2 extinguisher were fired into an oxygen consumption calorimeter. Based on a lack of oxygen and the presence of high quantities of CO_2, the instrument would respond with a high heat release reading.

[11] Estimates based on Exponent examinations of 18650 cells from various cell manufacturers.

electrolyte. The variability of cell case materials (nickel coated steel compared to aluminum foil) makes using weight percentage comparisons problematic. Use of these estimates for large format cells may be very inappropriate since internal designs are much more variable, particularly since some large format cell designs contain appreciable free electrolyte.

Thus, a pallet of ten thousand 2 Ah 18650 cells (approximately 20,000 Ah of capacity) would contain approximately 2–3 GJ of energy that could be released by the cells alone. Packaging materials would increase the total energy available for release.

The effect of packaging materials can be significant, for example, Exponent found that with notebook computers containing lithium-ion battery packs, the plastics associated with the notebook computers and various packaging materials would likely dominate expected heat release in a fire, and that the contribution of the lithium-ion cells might be negligible. Exponent[12] estimated the energy content of a battery containing 18650 cells by considering the total chemical energy that would be available from complete combustion of the electrolyte and the plastic separator and other insulators in each cell multiplied by the number of cells in the pack, as well as the electrical energy stored in the cells. This energy content was compared with the energy content of a notebook computer and all of its typical packaging when shipped as an individual unit. As part of this study, notebook computers were purchased on-line from Amazon.com. The computers and their packaging was disassembled and weighed. Data from the literature was used to estimate heat of combustion of component parts. To estimate the electrical energy content of the battery, the nominal pack capacity was assumed (assume 100% SOC), as listed by the manufacturer, to be fully converted into heat energy. This represented a conservative estimate, since batteries are most often shipped at 50% SOC and therefore contain approximately only half the electrical energy. Exponent estimated that batteries containing six cells contained total energy of approximately 3,500 kJ, while an associated notebook computer and its packaging contained chemical energy of approximately 60,000 kJ. The analysis indicated that the energy content of the batteries was less than 10% of the overall energy content of the notebook computers packaged for shipment.

A wide variety of consumer devices are shipped with lithium-ion batteries contained in or packed with the equipment and thus represent scenarios where the total energy of the package is likely to be dominated by the construction of the equipment being shipped and its packaging materials, rather than by the contribution of lithium-ion cells within the package. Shippable packages of consumer products such as power tools, cell phones, and DVD players that include lithium-ion batteries contained in or packed with equipment are similar to those used to ship notebook computers. A shippable package containing any of these items

---

[12] Harmon J, Gopalakrishnan P, Mikolajczak C, "US FAA-style flammability assessment of lithium-ion batteries packed with and contained in equipment (UN3481)," US Government Docket ID: PHMSA-2009-00095-0117, PHMSA-2009-00095-0119.1, PHMSA-2009-00095-0119.2, and PHMSA-2009-00095-0120.1, March 2010.

usually includes a sufficiently large and sturdy cardboard box to enclose and protect the device itself from damage, associated accessories, user manuals, inner packaging materials, and cushioning materials (packing peanuts, Styrofoam, paper, bubble wrap, etc.). The size and quantity of packaging materials will roughly scale with the size of the device and its battery: a small smart-phone package will contain a small battery likely consisting of a small single cell, while a larger power tool package will likely contain a larger multi-cell battery. The exact ratio of energy contained in the battery (electrical and chemical) to the total energy of the package will vary. However, that ratio it is likely to be low since the bulk of heat release will be produced by combustion of packaging materials.

Large format battery packs containing large cells or high cell counts will likely have total energy contents closer to bulk shipments of cells rather than bulk shipments of consumer electronics devices.

Fire Behavior of Cells and Battery Packs

As discussed above, there is no publically available data from large-scale lithium-ion cell or battery pack fire/fire suppression tests. There are a number of reasons for the lack of large-scale test data. The lithium-ion cell industry has been evolving rapidly, so there has been an inherent difficulty in defining an "average" cell, battery pack, or device. Thus, if testing were to be conducted and considered reasonably comprehensive, it would require testing of multiple models of cells, packs, or devices from multiple suppliers, and even so might quickly become obsolete as cell chemistries and mechanical designs evolved. Until recently there was very little cell or battery production within the US, resulting in limited interest in testing within the US. Cell thermal runaway incidents are rare, and a number of factors reduced the risk of fire incidents: most cells or packs entering the US were small, were packaged in or with equipment, and were shipped into the US at reduced state of charge. In the air transport sector, the risks associated with lith-ium-ion battery fires were considered significant enough to drive a test program. Thus, testing appropriate to scenarios encountered in air transport has been con-ducted, and since the testing was conducted to inform proposed regulations, the results of that testing are publically available.

In 2004, Harry Webster at the FAA published a report entitled, "Flammability Assessment of Bulk-Packed, Nonrechargeable Lithium Primary Batteries in Transport Category Aircraft."[13] This report described testing fire behavior of lithium primary cells in a 64-cubic-foot test chamber. The methodology described in this FAA test is designed to examine fire behavior under limited airflow con-ditions, specifically conditions that might be obtained within an aircraft cargo

---

[13] Webster H, "Flammability Assessment of Bulk-Packed, Nonrechargeable Lithium Primary Batteries in Transport Category Aircraft," DOT/FAA/AR-04/26, June 2004.

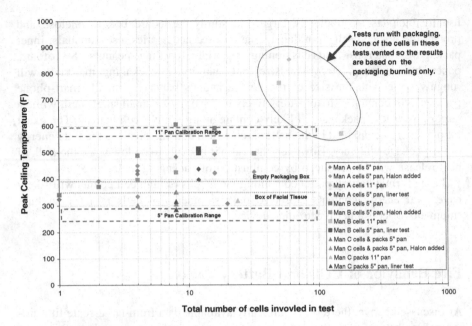

Fig. 6.1 Summary of peak ceiling temperatures for all tests conducted. (Mikolajczak CJ, Wagner-Jauregg A, "US FAA-Style Flammability Assessment of Lithium Ion Cells and Battery Packs in Aircraft Cargo Holds," Exponent Failure Analysis Associates, Inc., April 2005; PHMSA-RSPA-2004-19886-0044)

hold, or an aircraft unit load device. Under these conditions, airflow is very limited, the size and duration of an initiating fire is limited, combustion products and any added suppressants remain largely contained within the test chamber, and any heat generated by reactions remains largely contained within the test chamber. The techniques described in this FAA report have since been used by researchers at the FAA and at Exponent to examine the fire behavior of a variety of lithium-ion cells, battery packs, and consumer electronic devices containing or packaged with lithium-ion battery packs. Although these tests were conducted to address a specific air transport scenario: a cargo hold fire impinges on a shipment of cells or battery packs, they provide the most comprehensive temperature, heat flux, and fire behavior data publically available regarding lithium ion cell and battery pack fires, and therefore the results of these tests are discussed below.

In 2004, Exponent conducted FAA-style flame attack tests on single, multiple, and bulk packaged lithium-ion cells and battery packs.[14] Lithium-ion 18650 cells and battery packs containing 18650 cells at 50% SOC from three different manufacturers were tested in a 64-cubic-foot chamber. Calibration tests were run with

---

[14] Mikolajczak CJ, Wagner-Jauregg A, "US FAA-Style Flammability Assessment of Lithium Ion Cells and Battery Packs in Aircraft Cargo Holds," Exponent Failure Analysis Associates, Inc., April 2005; PHMSA-RSPA-2004-19886-0044.

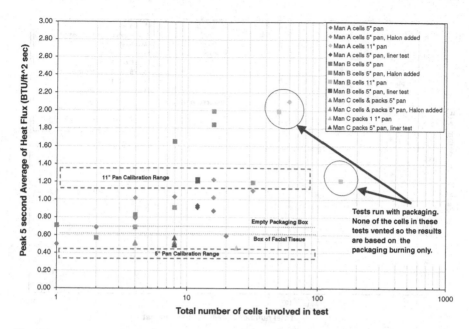

Fig. 6.2 Summary of peak 5-s averaged heat flux at the ceiling for all tests conducted. (Mikolajczak CJ, Wagner-Jauregg A, "US FAA-Style Flammability Assessment of Lithium Ion Cells and Battery Packs in Aircraft Cargo Holds," Exponent Failure Analysis Associates, Inc., April 2005; PHMSA-RSPA-2004-19886-0044)

both 5-inch and 11-inch fire pans. For comparison purposes, a test was run with empty bulk shipment packaging, and a test was run with a common box of facial tissue. Figure 6.1 shows the measured peak ceiling temperatures from all tests conducted as a function of the number of cells involved in each test. This figure shows that peak ceiling temperatures for most of the tests stayed within the range of peak temperatures measured for the 5-inch and 11-inch fire pans alone (calibration tests). Peak ceiling temperature did exceed the range of those produced by the 11-inch fire pan alone in tests of 50, 60, and 150 cells in bulk shipment packaging. However, it is important to note that no cells vented in these tests, so the energy released in these tests was from the 11-inch fire pan and the burning packaging material only. Figure 6.2 shows the peak 5-s average of the heat flux measured at the ceiling of the chamber: the highest value measured for bare cells was 2.0 BTU/ ft^2-s (23 kJ/m^2-s). Figure 6.3 shows the peak temperature measured approximately 12 inches above the chamber floor. All results except for the 150 cells in packaging test[15] cluster in the range of 1,000–1,400°F (540–760°C). These temperatures are consistent with those produced by burning normal combustibles such as packaging materials.

---

[15] The box arrangement in this test may have affected the reading at 12 inches.

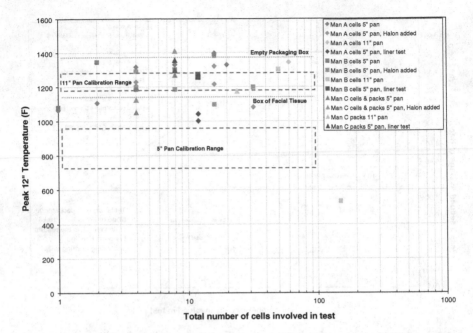

Fig. 6.3 Summary of peak temperatures measured 12 inches above the floor of the chamber. (Mikolajczak CJ, Wagner-Jauregg A, "US FAA-Style Flammability Assessment of Lithium Ion Cells and Battery Packs in Aircraft Cargo Holds," Exponent Failure Analysis Associates, Inc., April 2005; PHMSA-RSPA-2004-19886-0044)

In each bare cell test, all of the cells vented electrolyte and lost their external shrink-wrap. In many cases, the cell separator and carbon active material were also likely consumed. The resulting weight loss was approximately 7–10 g. A typical test proceeded as follows: the propanol was ignited, and flames impinged upon, and often surrounded the test sample. After 1–2 min, there was a series of audible "clicks" coinciding with small "puffs" of flames from the cells. These "clicks" and "puffs" were indications of preliminary vent releases, likely resulting from activation of cell charge interrupt devices. After approximately a 1-min delay, the main cell vents began to open, releasing jets of flammable vapor (electrolyte and plastic decomposition products) and producing a hissing sound. The vapors were ignited by the burning propanol and resulted in jetting flames emanating from the cells for times on the order of a few seconds. In a few tests, some cells ruptured their cases and expelled their contents.

Figure 6.4 shows all of the ceiling and 12-inch temperature measurement data overlaid for this series of tests. It is evident that the ceiling temperature measurements generally fall within a narrow band. The maximum measured ceiling temperature for these tests was 581°F (305°C). The 12-inch temperature measurements are more variable, reflecting the thermocouple proximity to flames. The maximum measured temperature at 12-inches above the floor was 1,390°F (754°C).

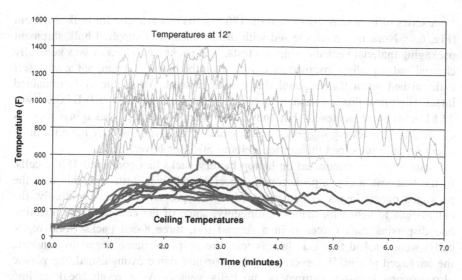

Fig. 6.4 Compilation of temperature data for all bare cell tests. (Mikolajczak CJ, Wagner-Jauregg A, "US FAA-Style Flammability Assessment of Lithium Ion Cells and Battery Packs in Aircraft Cargo Holds," Exponent Failure Analysis Associates, Inc., April 2005; PHMSA-RSPA-2004-19886-0044)

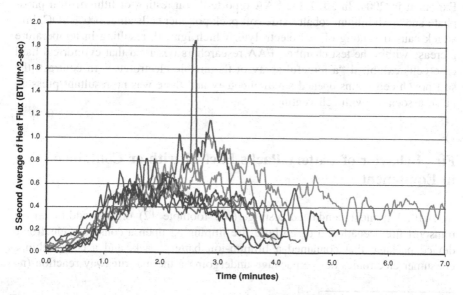

Fig. 6.5 Compilation of 5-s averaged heat flux at the ceiling for all bare cell tests. (Mikolajczak CJ, Wagner-Jauregg A, "US FAA-Style Flammability Assessment of Lithium Ion Cells and Battery Packs in Aircraft Cargo Holds," Exponent Failure Analysis Associates, Inc., April 2005; PHMSA-RSPA-2004-19886-0044)

A series of tests were conducted on 18650 cells as packaged for bulk shipment (Fig. 6.5). None of the cells tested with the manufacturer-supplied bulk shipment packaging material vented during the tests. The packaging material was generally charred and partially consumed by flames from the initial fuel pan, but always self extinguished when the propanol flame extinguished. All of the cells remained intact. High heat fluxes measured at the chamber ceiling (up to 2.10 BTU/ft^2-s or 24 kJ/m^2-s) were achieved, but they were due to burning packaging such as cardboard. This testing suggested that a small, short-lived fire may have minimal effect on bulk packaged lithium-ion cells at 50% (or lower) SOC.

Tests were also conducted on laptop battery packs that contained 18650 cells. In the first test, a single pack containing eight cells was tested. In this test, the packaged plastic began to be consumed by the propanol flame. Eventually, the cells began to vent with flames. Ultimately, some of the cells ruptured, ejecting and dispersing their contents. In a second test, three 8-cell packs (total of 24 cells) were stacked together. In this test, the propanol flame began to consume the packaged plastic. However, once the propanol flame extinguished, the plastic also stopped burning. Ultimately, no cells vented. As a result, peak ceiling temperatures and average heat flux were lower for the 3-pack test than for the single pack test.

In 2006, FAA conducted similar tests using 18650 lithium-ion cells at 50% SOC and 100% SOC.[16] They observed behavior to similar to that reported by Exponent in 2004. In 2010, the FAA reported[17] on testing of lithium iron phosphate and 8-Ah lithium cobalt oxide soft pack polymer cells all at 100% SOC. Fire attack caused venting of cell electrolyte, which ignited, resulting in temperature increase within the test chamber. FAA researchers observed that cylindrical hard case cells exhibited "a forceful spray of flammable electrolyte." In contrast, the soft pouch cell seams opened simultaneously and there was no resultant pressure pulse associated with cell venting.

Fire Behavior of Battery Packs Packed with or Contained in Equipment

In 2010, Exponent conducted testing, (see footnote 12) of two additional air transport fire scenarios: a cargo hold fire impinging upon a consumer electronic device package that contained a lithium-ion battery pack, and a cell within a consumer electronics device package undergoing a thermal runaway reaction (no

---

[16] Webster H, "Flammability Assessment of Bulk-Packed, Rechargeable Lithium-Ion Cells in Transport Category Aircraft," DOT/FAA/AR-06/38, September 2006, http://www.fire.tc.faa.gov/pdf/06-38.pdf.

[17] Summer SM, "Flammability Assessment of Lithium-Ion and Lithium-Ion Polymer Battery Cells Designated for Aircraft Power Usage," DOT/FAA/AR-09/55, January 2010, http://www.fire.tc.faa.gov/pdf/09-55.pdf.

external heating). Both of these scenarios were conducted under the limited airflow conditions appropriate to aircraft transport scenarios. The observations from of these tests were consistent with reported air cargo incidents involving lithium-ion batteries described in Chap. 5. However, the results also provided insights that are applicable to a broader range of conditions.

Flame attack testing on packaged consumer electronics devices showed that initial fire development was dominated by packaging materials—a result that is independent of the airflow conditions. Flames initially attacked and ignited external packaging materials (cardboard), and were likely to be self extinguished due to limited airflow before cells vented and entered thermal runaway. Fire attack testing on systems with and without batteries showed that the temperature and heat flux values from the initial flaming period were effectively identical (within normally expected variation of fire tests). This indicated that the presence of the battery had no discernable effect on the overall heat release during this time.

This testing also demonstrated a unique hazard associated with fires involving lithium-ion cells and battery packs: that without sufficient cooling, cell thermal runaway reactions can occur significantly after flame suppression—a result that is also independent of the airflow conditions. No cooling was provided to the test articles (suppression by smothering of the fire). Recall that because the testing was conducted within a small enclosure, heat from the initial fire was retained in the area of the test articles. Thus, even though cells did not vent during the initial fire, they did ultimately undergo venting, often after an extended period. The time to venting depended on a variety of factors such as chamber temperature, airflow conditions, battery location, cell design, and cell SOC. The shortest time to cell venting observed in testing (approximately 19 min) involved fire impinging on the region directly adjacent to cells packed with equipment. Longer times to venting were likely if the battery was contained in equipment, or flames did not directly impinge upon the area of the battery. Exponent observed that subsequent cell thermal runaway and venting (after the initial period) could produce hot spots that caused re-ignition of combustibles, if sufficient air can enter the enclosure to sustain flaming combustion.

These observations may have significant implications on fire fighting procedures, specifically fire protection and fighting strategies, fire scene overhaul procedures, and fire scene monitoring for rekindles. Specifically, if a fire occurs adjacent to stored lithium-ion cells and battery packs, those cells and battery packs must be protected from relatively modest (compared to flashover) overheating, or cells may begin to vent and ignite, spreading the fire more rapidly than would be expected for normal combustibles. On fire scenes where large quantities of lithium-ion cells have been involved, decisions regarding overhaul procedures must be made with an understanding that as cells are uncovered, moved, or damaged (crushed/punctured) by overhaul procedures, they may undergo thermal runaway reactions and vent, they may ignite, and they may generate, or themselves might become projectiles. Similarly, the potential for rekindles will be high at such fire scenes, and these scenes will require extended monitoring.

Some important cell initiation testing results were not affected by limited airflow to the test chamber. They demonstrated that cell thermal runaway events within packaged consumer electronics devices were unlikely to propagate beyond the packages due to low shipping states of charge and airflow limitations imposed by packaging materials. For these tests, single cells within battery packs were fully charged and connected to heaters that could induce thermal runaway reactions when energized. This initiation method was developed to mimic a severe cell internal fault. The battery packs were packaged as for shipment with their associated electronic equipment (in this case, notebook computers). Using the installed heater, a single fully charged cell within the pack was driven to thermal runway. Cell venting produced soot and smoke that in some cases escaped the packaging and became visible to observers. If the non-initiating cells within the battery pack were at a reduced SOC, flaming combustion was unlikely to occur due to limited oxygen within the packaging material. Note that no ignition source was present external to the device package. If cells were near 100% SOC, ignition of vent gases might occur. Because testing was conducted in an FAA enclosure with limited airflow, when flaming combustion was initiated the resulting fire was a short duration and relatively low intensity event, and the fire self-extinguished due to limited oxygen in the chamber.

Effectiveness of Suppressants

Fires involving lithium-ion cells are the result of electrolyte burning, which is a hydrocarbon/air flame. Thus, many flame suppression agents will be effective in suppressing flaming combustion. However due to the electrical nature of battery packs, particularly the high voltages associated with large format battery packs, conductive suppression agents may not be a good choice. In addition, because of the potential for re-ignition due to cascading cell thermal runaway reactions, an ideal suppressing agent will stay suspended and prevent re-light of combustible mixture from cell hot surfaces. Suppressants shown to be effective include: inert gas/smothering of flames[18] (fire behavior testing data (see footnote 14) indicates that smothering is effective in preventing flaming, but will not cool cells and prevent thermal runaway propagation), carbon dioxide (Exponent typically uses carbon dioxide extinguishers to suppress flaming of cells during testing—this will not cool cells and prevent thermal runaway propagation), water (a number of sources see footnote 18,[19,20] have described the effectiveness of water to suppress flaming and cool cells), and Halon (see footnotes 18 and 20).

---

[18] Lain MJ, Teagle DA, Cullen J, Dass V, "Dealing with In-Flight Lithium Battery Fires in Portable Electronic Devices," CAA Paper 2003/4, July 30, 2003.

[19] Advance Change Notices to NSTM 555VIR12 and NSTM 555V2R11 for Lithium Battery Firefighting Procedures, July 21, 2009.

[20] http://www.fire.tc.faa.gov/systems/handheld/handheld.asp, include a link to a video entitled, "Extinguishing In-flight Laptop Computer Fires."

There is limited published data regarding the selection of suppressants for use on lithium-ion battery fires. The design of suppression systems in battery manufacturing facilities is generally considered proprietary information and is not publically available. Testing data that is available has been published is related to very specific lithium-ion battery applications, primarily the suppression of fires in air transport: fires that might occur in a passenger cabin, where very limited numbers of cells could become involved and Halon extinguishers and water are available suppressants, (see footnote 18) and fires that might occur in aircraft cargo holds, where Halon is the available suppressant. Full scale fire suppression testing is necessary to evaluate specific storage configurations, quantities, arrangements and fire suppression system design criteria and overall effectiveness.

Navy Sea Systems Command released an Advanced Change Notice for Lithium Battery Firefighting Procedures (see footnote 19). In this document, the Navy recommends (based on limited testing), the use of "a narrow-angle fog of water or AFFF" to cool batteries, suppress "fireballs," and reduce the likelihood of thermal runaway propagation.

The FAA studied suppression of lithium-ion batteries with water and Halon 1211, as these are typically available in hand extinguishers aboard commercial aircraft (see footnote 20). As a first choice, the FAA recommends the use of water to suppress fires involving notebook computers, because water will both extinguish flames and suppress thermal runaway propagation. As a second choice, the FAA recommends using Halon 1211 to knock down flames, followed by deluge from available water sources (such as bottles of drinking water). Halon 1211 alone will not prevent re-ignitions of cells due to propagation of cell thermal runaway reactions. In FAA tests, application of ice did not sufficiently cool cells to prevent thermal runaway propagation.

In 2010, the FAA reported on testing lithium iron phosphate and 8-Ah lithium cobalt oxide soft-pack polymer cells (see footnote 17). Halon 1211 was able to successfully extinguish flames from these cells. In addition, the iron phosphate cells did not continue to vent or re-ignite once the Halon 1211 was applied. However, Halon 1211 was not able to suppress re-ignition of the soft-pouch polymer cells (cobalt oxide chemistry).

Halon 1301 is the least toxic of the Halon fire suppressants and is considered to have superior fire extinguishing characteristics. In particular, it rapidly knocks down flaming combustion, has a penetrating vapor that can flow around baffles and obstructions, leaves no residue, is non-corrosive, requires small storage volumes, is non-conductive, and is colorless, which prevents the generation of false fire alarms by obscuration. Halon does not act by displacing oxygen, rather it acts by interfering with the chemistry of combustion, specifically by terminating chain branching reactions that occur in the gas phase in typical hydrocarbon/air flames. The fact that Halon is effective in suppressing lithium-ion battery flames is another indication that these flames are substantially similar to typical hydrocarbon/air flames. Halon 1301 (bromotrifluoromethane) is a methane derivative. The bromine atom confers strong fire suppressant properties,

Table 6.4 Minimum required and design volume percentage of Halon 1301 at 25°C (77°F) that will prevent burning of various vapors

Fuel	Volume % Halon 1301 in air required for flame extinguishment[a]	Design concentrations for flame extinguishments (volume % Halon)[b]
Methane	3.1	5.0
Propane	4.3	5.2
n-Heptane	4.1	5.0
Ethylene	6.8	8.2
Acetone	3.3	5.0
Benzene	3.3	5.0
Ethanol	3.8	5.0
Plastics	4–6	

[a] Taylor GM, "Halogenated Agents and Systems," Section 6, Chapter 18, *Fire Protection Handbook*, 18th ed., National Fire Protection Association, 1997
[b] Grant CC, "Halon Design Calculations," Section 4, Chapter 6, *SFPE Handbook of Fire Protection Engineering*, 2nd ed., Society of Fire Protection Engineers, 1995

while the fluorine atoms confer stability to the molecule and reduce its toxicity. Bromine atoms interfere with the free radical and chain branching reactions that are important in combustion.

Halon 1301 is generally considered very effective for electrical fires (Class C fires),[21] flammable liquid and gas fires (Class B fires),[22] and surface-burning flammable solid (such as thermoplastic) fires. However, Halon 1301 has minimal effectiveness on reactive metals, rapid oxidizers, and deep-seated Class A fires.[23] Halon 1301 is minimally effective on deep-seated Class A fires because it works by interfering with the chemical reactions that create flames; it does not cool the fuel feeding the fire. Thus, while Halon 1301 can extinguish the flaming portion of a Class A fire, the glowing deep-seated portion of the fire can continue to smolder and spread at a reduced rate.

The strong effect of Halon addition can be seen upon examining the flammability limits of fuel/air/Halon mixtures, and comparing them with the flammability limits of fuel/air/inert diluent mixtures. When small quantities of Halon are added to a fuel/air mixture, they narrow the range in which that mixture is flammable.[24] Halon is far more effective at narrowing the flammable range than an inert diluent. If sufficient Halon is added, the flammable range of a mixture, even at an elevated

[21] NFPA 10, "Standard for Portable Fire Extinguishers," defines Class C fires as fires involving energized electrical equipment.
[22] NFPA 10, "Standard for Portable Fire Extinguishers," defines Class B fires as fires involving flammable liquids and gases.
[23] NFPA 10, "Standard for Portable Fire Extinguishers," defines Class A fires as fires involving ordinary combustible materials such as paper, wood, cloth, and many plastics.
[24] See for example, the effect of Halon addition on flammability of methane in "Basics of Fire and Science," Section 1, Chapter 1, *Fire Protection Handbook*, 18th Edition, National Fire Protection Association, 1997.

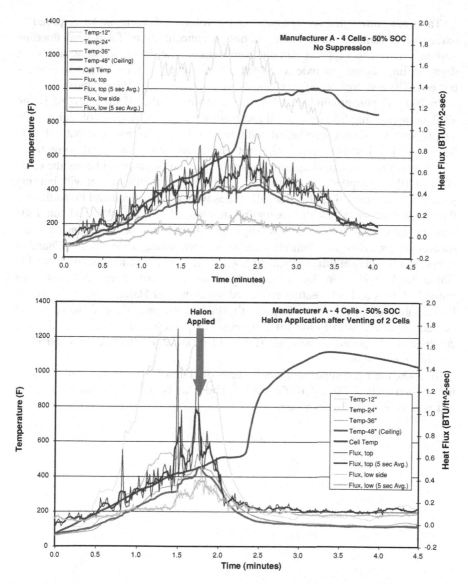

Fig. 6.6 Tests with four Manufacturer A cells, without suppression (*top*), and with Halon 1301 application after cells began to vent (*bottom*). (Mikolajczak CJ, Wagner-Jauregg A, "US FAA-Style Flammability Assessment of Lithium Ion Cells and Battery Packs in Aircraft Cargo Holds," Exponent Failure Analysis Associates, Inc., April 2005; PHMSA-RSPA-2004-19886-0044)

temperature, is eliminated and the mixture cannot be ignited. Note that production of Halon was banned by the Montreal Protocols, as this material contributes to the destruction of the ozone layer. Halon in use today is from recycled sources only, primarily for protection of aircraft.

Table 6.4 shows the average percent by volume of agent in air required to extinguish a flame. It also shows the design concentrations for a total flooding system required to suppress flaming combustion. The design concentrations for flame extinguishment include an added safety factor over the required concentrations. These design recommendations are approximately 5% for most fuels.

In 2004, Exponent conducted FAA-style testing of the effectiveness of Halon 1301 in suppressing lithium-ion cell and battery pack fires. A series of Halon 1301 suppression tests were conducted with bare 18650 cells and computer laptop battery packs. The bare cells were not electrically connected, but were taped together. Ignition was accomplished by igniting a pan of propanol below the cells. Halon 1301 was applied late in each test, once cells had begun to vent with burning jets. Within seconds of application, all flames were extinguished and no additional flaming was observed for the continuing duration of the test. When Halon 1301 was applied there was a precipitous drop in the chamber temperatures and heat flux measurements. This was entirely consistent with flame suppression. Chamber temperatures and heat fluxes remained low for the duration of the testing. Note that Halon 1301 application did not cool the cells (Fig. 6.6). Thermal runaway of individual cells and cell venting continued to occur after Halon 1301 was applied. Examination of all cells from Exponent's Halon 1301 tests showed that they had vented. However, with Halon 1301 present, this process did not result in flaming combustion. Exponent concluded Halon 1301 is very effective in controlling burning lithium-ion cells.

In 2006, the FAA conducted similar tests of Halon 1301 suppression on 18650 lithium-ion cells at 50 and 100% SOC.[25] The FAA observed similar behavior to test results reported by Exponent.

---

[25] Webster H, "Flammability Assessment of Bulk-Packed, Rechargeable Lithium-Ion Cells in Transport Category Aircraft," DOT/FAA/AR-06/38, September 2006, http://www.fire.tc.faa.gov/pdf/06-38.pdf.

Chapter 7
Lithium-Ion Fire Hazard Gap Analysis

There are a number of gaps in the available information regarding lithium-ion fire hazards related to designing fire protection systems to protect personnel working in facilities where lithium-ion cells are stored and to protect personnel responding to a potential incident, as well as for protecting the surrounding environment in the case of an incident, and finally to protect property and structures. These gaps can be generally categorized into three areas:

- A limited understanding of the composition and flammability of leaked cell electrolyte and cell vent gases
- The lack of a fire protection commodity specification for bulk packaged lithium-ion cells or large format lithium-ion battery packs
- Limited data regarding the effectiveness of potential suppressants, specifically water

we discuss each of these areas in further detail as well as fire and flammability testing approaches that could be used to close these gaps.

Leaked Electrolyte and Vent Gas Composition: Gap 1

Although Sandia has published some data regarding leaked electrolyte and vent gas composition, there are a number of significant gaps in the data that are important for detection of hazardous events, protection of affected personnel, and mitigation of fire and explosion hazards.

Gap 1.1: The Sandia data provides insight regarding gas composition that can be used to detect cell leakage or vent gas products in a facility and initiate response procedures. However, there is no data currently available to recommend a particular sensor or sensor package for detecting leaking or venting cells. Small-scale, single-cell tests could be used to assess the effectiveness of various sensor

C. Mikolajczak et al., *Lithium-Ion Batteries Hazard and Use Assessment*,
SpringerBriefs in Fire, DOI: 10.1007/978-1-4614-3486-3_7,
© Fire Protection Research Foundation 2011

packages to a wide range of cell models, particularly to test effectiveness when vent gases do not ignite. Sensor packages could then be tested and evaluated for effectiveness in detecting fires involving lithium-ion cells either in small-scale tests and validated in conjunction with full scale tests.

Gap 1.2: The Sandia data identifies the major components of leaked and vented cell gases. However, cell vent gas toxicity could be more strongly dependent upon minor gas components such as fluorinated compounds for example HF, COF_2, and F_2. Acceptable alarm or evacuation threshold levels are therefore currently not well defined. The first step to addressing this gap would be to conduct small-scale single cell testing of a variety of cell chemistries to more comprehensively evaluate cell leakage and vent gas compositions, including vent gas post combustion products. Gas sampling during full scale testing could be conducted to assess gas compositions during fires. Following this evaluation, a comparative assessment of the hazards associated with identified gas components can be made, and recommendations for warning, alarm, and evacuation levels can be generated.

Gap 1.3: Testing and experience discussed previously have shown that cell vent gases are ignitable, and thus release of these gases could pose a deflagration hazard. However, the composition of these gases suggests that their flammability limits may be fairly narrow. Determining the flammable range of cell vent gases could improve hazard mitigation approaches such as gas exhaust handling. This could be particularly applicable to cell manufacture facilities where cells must undergo formation, a process that results in gas generation. Single cell testing could be used to collect gas samples for flammability testing.

Lithium-Ion Cell and Battery Commodity Specification: Gap 2

Gap 2.1: At present there is no fire protection commodity classification for lithium-ion cells. At present, there is no publicly available large-scale fire test data for lithium-ion cells that can be used to fully assess the storage hazards of lithium-ion cells or batteries or to determine an appropriate commodity classification that could be used to provide an overall fire protection suppression strategy. Developing an appropriate commodity classification for lithium-ion cells and batteries can involve characterizing, as appropriate: ignitability, total heat release, heat release rate, time to peak heat release rate, thermal profiles within a test sample, burning duration, energy density, the composition of gas phase combustion and pyrolysis products, the reactivity of the fuel load to suppression agents, effectiveness of suppression applications and the way in which these devices are arranged, and the storage configuration, quantity, geometry, and arrangement of the commodity.

Commodity classifications for water based suppression strategies are described in NFPA 13, *Standard for the Installation of Sprinkler Systems*,[1] which addresses sprinkler systems applications and proposes requirements for storage protection (the question of whether sprinkler systems are appropriate will be discussed below). The commodity classification relates directly to the fire protection system design requirements. According to Section 5.6.1.1.1, "Commodity classification and the corresponding protection requirements shall be determined based on the makeup of individual storage units (i.e., unit load, pallet load)." Commodities are typically classified as one of the following (§5.6.3.1) listed classes, or they may be determined to be a special hazard that requires special consideration (i.e., tires, flammable liquids, aerosols, etc.):

Class I—a noncombustible product that meets one of the following criteria: product is placed directly on wood pallets; product is placed in single-layer corrugated cartons, with or without single-thickness cardboard dividers, with or without pallets; product is shrink-wrapped or paper-wrapped as a unit load with or without pallets.

Class II—a noncombustible product that is in slatted wooden crates, solid wood boxes, multiple-layered corrugated cartons, or equivalent combustible packaging material, with or without pallets (§5.6.3.2).

Class III—a product fashioned from wood, paper, natural fibers, or Group C plastics with or without cartons, boxes, or crates and with or without pallets. A Class III commodity shall be permitted to contain a limited amount (5 wt% or volume or less) of Group A or Group B plastics (§5.6.3.3).

Class IV—a product, with or without pallets, that meets one of the following criteria: product constructed partially or totally of Group B plastics; product consists of free-flowing Group A plastic materials; product contains within itself or its packaging an appreciable amount (5–15 wt% or 5–25 vol%) of Group A plastics.

Group A, B, and C Plastics—Plastics, elastomers and rubbers. If the material to be stored contains plastic, elastomer, or rubber, a group classification (A, B, or C) is also determined according to its composition in order to determine the protection system requirements. The specific Group depends on the type of plastic, elastomer or rubber (§5.6.4).

NFPA 13 provides a list of commodity classes for various commodities in Table A.5.6.3. Different types of batteries and the recommended commodity classification for those batteries are mentioned:

---

[1] The most widely used standard in terms of design and installation of sprinkler systems is NFPA 13 *Standard for the Installation of Sprinkler Systems* developed by the National Fire Protection Association. It addresses sprinkler systems applications and proposes requirements for storage protection. Most of the current sprinkler system design criteria are based on full-scale testing and the application of experimental results that prove to provide a minimum level of protection. The most current edition of NFPA 13 is the 2010 edition.

- Dry cells (non-lithium or similar exotic metals) packaged in cartons: Class I (for example alkaline cells);
- Dry cells (non-lithium or similar exotic metals) blister packed in cartons: Class II (for example alkaline cells);
- Automobile batteries—filled: Class I (typically lead acid batteries with water-based electrolyte); and
- Truck or larger batteries, empty or filled Group A Plastics (typically lead acid batteries with water-based electrolyte).

NFPA 13 currently does not provide a specific recommendation for the commodity classification (or fire protection strategies) for lithium-ion cells or complete batteries containing several cells. According to the NFPA Automatic Sprinkler System Handbook[2]: "Classification of actual commodities is primarily based on comparing the commodity to the definitions for the various commodity classes."

Lithium-ion cells and batteries might be compared to truck or larger batteries. However, a number of features specific to lithium-ion batteries could make this classification inaccurate and the recommended fire suppression may not be appropriate:

- Flammable versus aqueous electrolyte.
- The potential to eject electrodes/case material (projectiles) upon thermal runaway.
- Latency of thermal runaway reactions (cell venting can occur sequentially and after a significant delay resulting in re-ignition of materials).
- Large format battery packs may exhibit voltages much higher than typical truck batteries.
- Individual cells generally have metal versus plastic outer shells.

The venting and projectile potential for the lithium-ion cells might make them comparable with aerosol products, which typically utilize a flammable propellant such as propane, butane, dimethyl ether, and methyl ethyl ether. However, these products generally do not have associated electrical energy and are not as susceptible to re-ignition events. As they contain flammable electrolyte, lithium-ion cells might also be compared to commodities such as ammunition or butane lighters in blister packed cartons (high energy density).

For commodities not specifically covered by NFPA 13, full-scale fire suppression tests are typically used to determine the commodity classification for that specific commodity. Indeed, most current sprinkler system design criteria are based on classifications of occupancies or commodities that have been developed from the results of full-scale fire suppression test data and the application of

---

[2] Dubay C (ed.), *Automatic Sprinkler System Handbook*; §5.6.1 Commentary, National Fire Protection Association, Quincy, MA, 2010.

experimental results that have been shown to provide a minimum level of protection. According to the NFPA Automatic Sprinkler System Handbook[3]:

> Where commodities are not currently defined, commodity classification testing can provide an accurate comparison between the proposed commodity and known commodity classifications. This testing is essential when determining acceptable sprinkler design criteria for new or unknown commodities where a meaningful comparison cannot be made between the given commodity and other known commodity classifications. Bench-scale testing is not useful for making precise commodity classifications.

One of the main reasons that specific test data are required when determining the commodity classification of a new or unknown commodity is that the current ability of an engineering analysis is incapable of defining sprinkler suppression characteristics.[4]

Annex C of NFPA 13 (2010 edition) provides a description of experimental procedures used to develop safety criteria and determine water density, rack arrangement and sprinkler characteristics requirements when existing knowledge is too limited. This reference should provide the basic framework for full scale testing. There are also a number of factors that must be considered specific to lithium-ion cell and battery pack testing discussed below.

Gap 2.1a: What is the appropriate commodity specification for bulk packaged lithium-ion cells? The bulk packed cell storage scenario is characterized by low packaging to battery volume ratios, and low cell voltages. This scenario is consistent with many cell storage situations during manufacturing, transport, and to some extent recycling. It also approximates the electric vehicle battery pack scenario with regard to energy density, but does not include high voltage effects. Thus, testing of bulk packaged lithium-ion cells should be the first priority for full scale fire testing aimed at determining acceptable suppression system design criteria. Addressing this gap will require full scale testing to create a benchmark. However as lithium-ion technology evolves, single cell or medium scale testing may be required to compare new cell designs with benchmark cells. Conducting full scale benchmark testing will require defining a number of parameters:

1. Cell chemistry: at present, commercial lithium-ion cells using cobalt-oxide based cathodes exhibit the highest energy densities (highest capacities in any given form factor) of cells on the market and are fairly common. Cells with higher energy densities are likely to produce the most severe reactions, thus Exponent recommends conducting full scale tests with high energy density (high capacity) cells.
2. Cell form factor: The 18650 cylindrical cell continues to be the most common lithium-ion cell form factor. Cells of this form factor are used in both consumer electronics devices, and electric vehicle applications. This cell form factor will

---

[3] Dubay C (ed.), *Automatic Sprinkler System Handbook*; §5.6.1 Commentary, National Fire Protection Association, Quincy, MA, 2010.

[4] Dubay C (ed), *Automatic Sprinkler System Handbook*; §5.6.1.1 Commentary, National Fire Protection Association, Quincy, MA, 2010.

provide a reasonable depiction of the fire behavior of many hard case cell designs, allowing assessment of high pressure cell venting and the potential for cells to become projectiles. Thus, Exponent recommends that 18650 cells be used for full scale testing. However, full scale testing of soft-pouch polymer cells should also be a high priority. Unlike hard case cells, soft-pouch cells will not exhibit high pressure venting or projectile behavior, however, they are typically shipped in plastic trays that may affect fire growth significantly.

3. Cell internal separator: Most cells on the market contain polymeric shut-down separators that melt at thermal runaway temperatures and can contribute flammable degradation products to fires. Exponent recommends that cells with these types of separators be used for testing.

4. Cell state of charge: Bulk packaged cells are typically shipped and stored at reduced states of charge (50% or below) to prevent degradation and aging. Thus, for most storage scenarios, testing cells at reduced states of charge would be appropriate. However, it is possible for manufacturers to ship fully charged cells. Thus, to capture the most severe possible behavior, Exponent recommends conducting testing of cells at 100% SOC. However testing at 50% SOC or below should be conducted for comparison purposes and to evaluate the overall fire protection strategy and corresponding suppression system design criteria.

5. Packaging configuration: Bulk packaged cells are typically shipped and stored in palletized boxes that may be shrink-wrapped together. Exact box dimensions and layout depend upon individual cell manufacturers. Exponent recommends that initial full scale tests be conducted on single pallets of cells. Once fire behavior has been documented, multiple pallet tests can be attempted to assess fire suppression system effectiveness and appropriate design criteria as well as storage arrangement and geometries.

6. Cell age: Used cell thermal runaway behavior may be different from new cell behavior depending upon cell usage history. However, there would be considerable difficulty in controlling test parameters for batches of used cells collected for recycling and thus Exponent recommends that initial testing be conducted with new cells.

7. Initiating event: There are two likely fire initiation scenarios that should be considered: that of a fire in a facility impinging on stored lithium-ion cells and that of a lithium ion cell spontaneously undergoing a thermal runaway reaction that causes ignition of cells. For new bulk packaged cells the external fire impingement scenario is likely the best choice for initial testing as this scenario is likely to result in the most severe result regardless of cell chemistry, state of charge, or packaging details: the impinging fire will provide heating of multiple cells simultaneously, and a competent ignition source for cell vent gases. For suppression tests, suppressant application can be delayed until evidence of well established thermal runaway propagation has been verified.

It is impractical to conduct full scale testing on multiple cell chemistries, multiple cell form factors, multiple states of charge, multiple packaging arrangements, etc. It is also impractical to conduct full scale testing of all new cell designs that might involve higher cell capacities or energy densities. Thus, in order to aid determination of whether a commodity specification developed based on benchmark testing is relevant to a different lithium-ion cell type or packing arrangement, a small-scale testing program should be developed to allow comparison between cells subjected to full scale testing, and cells of other types (other chemistries, form factors, states of charge, etc.). Single cell testing could be used to assess a number of fire behavior characteristics including:

- Vented electrolyte composition and volume for a broad range of cell states of charge.
- Vented electrolyte flammability limits.
- Heat release rates of cells undergoing thermal runaway at various conditions (this would require modification of traditional oxygen consumption calorimetry techniques).

  - New cells versus aged cells that have reduced capacity but also may have accumulated dead lithium that could affect heat release rates during thermal runaway.
  - Cell chemistry and cell state of charge have a significant effect on initial heat release rate (as has been demonstrated by ARC testing, which is generally terminated before ignition of vent gases occurs), but cell electrolyte content may dominate total and peak heat release rates during fires.

Cells of a range of chemistries, form factors, and states of charge could then be compared with cells subjected to full scale tests on the basis of potential for producing flammable vent gases, total heat release, and heat release rates. The small scale testing program should allow fairly rapid assessment of new cell types to determine applicability of the commodity specification developed with full scale testing. Ideally, such a small scale testing program would be validated by multiple full scale tests.

Gap 2.1b: What is the appropriate commodity specification for large format lithium-ion batteries? This battery packaging scenario is characterized by low packaging to battery volume ratios just as the bulk packed cell case, but also by high battery pack voltages and the possibility of high current short circuits and electric arcing. The same considerations discussed under Gap 2.1a apply to testing large format battery packs. In addition, there is as of yet no standard battery pack or module configuration, and very few battery packs are currently being mass produced. Therefore, if full pack testing is to be attempted, obtaining battery packs to test may be the primary hurdle. As with testing of bulk packed new cells, initiation through fire impingement is likely to produce the most severe results, as this mode of initiation is likely to cause multiple cells to undergo thermal runaway near the beginning of the test, and to facilitate thermal runaway propagation by preheating many cells near the point of fire impingement. It is also experimentally the most convenient initiating event since it does not require that modifications be made to the large format battery pack. However, if the testing results are to be applied to locations where fully charged battery packs could be damaged such as crash test facilities or service locations, or locations where battery packs will be routinely charged, it may be appropriate to conduct tests where initiation is caused by thermal runaway of a single cell. This type of testing could provide insight on the fault development mechanisms for packs containing large cells or large parallel arrays of cells.

An approach similar to that described in Gap 2.1a may be useful in allowing comparisons between a variety of battery pack types with those subjected to full scale testing. Development of a small scale cell-level testing program may be appropriate.

Gap 2.1c: What is the commodity specification for lithium-ion cells contained in or packed with equipment? This battery packaging scenario is characterized by high packaging to battery volume ratios, low battery pack voltages, and small short-circuit currents. The commodity specification for packaged consumer electronics devices is currently based on the plastics content of the devices and their packages rather than the presence or absence of lithium-ion cells within the packages. Based on energy release analysis and fire attack testing of batteries contained in or packed with equipment described in Chap. 6: Lithium-Ion Fire Hazard Assessment, this classification is likely appropriate for many packaged goods. However, it is likely that at some lithium-ion cell density, a commodity classification based on device plastic content will be inappropriate. Full scale fire testing should be used to identify a cell density at which a bulk packaged lithium-ion cell commodity classification should be applied to packages of consumer electronics devices that contain lithium-ion cells.

Gap 2.2: Packaging details can affect the propensity for thermal runaway propagation. A bench scale test program involving relatively small arrays of cells could be developed to compare the effect of varying packaging approaches.

Suppressant Selection: Gap 3

The information available due to publically available testing conducted to date does not allow a comprehensive assessment of whether traditional water-based automatic sprinkler systems, water mist systems or some other water-based suppression system would be the most efficient for the protection of stored lithium ion cells or batteries. However, a number of sources including the FAA, and the US Navy recommend the use of water as a cooling and extinguishing agent. Water-based automatic sprinklers are the most widely used fire suppression equipment and have proved their efficiency and reliability over the years. Many locations are currently provided with the infrastructure necessary to facilitate suppression strategies using water based suppression systems. Therefore, based on current knowledge and infrastructure, a water-based fire suppression system is the strongest candidate for the protection of stored lithium ion cells and batteries.

Gap3.1: The sprinkler density and water flow rates required to suppress thermal runaway propagation of bulk packaged cells is unknown. Full scale fire and suppression testing will be required to address this question. Full scale tests would be required to assess the effectiveness of various suppression approaches including water sprinklers on full assembled pallets of cells.

Gap 3.2: The sprinkler density required to suppress thermal runaway propagation of large format battery packs that may pose an additional high voltage hazard is unknown. Full scale fire suppression tests would be requires to assess the effectiveness of sprinklers on large format battery packs.

Gap 3.3: The sprinkler density appropriate for suppressing fires involving packaged consumer electronics devices is currently based on the plastics content of the devices and their packages rather than the presence or absence of lithium-ion cells within the packages. Based on energy release analysis and fire attack testing of batteries contained in or packed with equipment described in Chap. 6: Lithium-Ion Fire Hazard Assessment, this classification is likely appropriate for many packaged goods. However, it is likely that at some lithium-ion cell density, a commodity classification based on device plastic content would be inappropriate. Full scale fire suppression testing could be used to identify a cell density at which a bulk packaged lithium-ion cell commodity classification should be applied.

Gap 3.4: The environmental contamination implications for using water based suppression is unknown. At present, no analysis has been conducted on the composition of run-off water used for suppressing fires involving lithium-ion cells. Run-off water could contain a wide range of potentially hazardous components including fluorinated compounds and metal oxides. Sampling and analysis of run-off water from full fire suppression scale testing should be conducted to asses the potential hazard to the environment.

Gap 3.5: Traditional water sprinkler system –based suppression may not be the most effective method for suppressing fires involving lithium-ion batteries. A number of other suppression approaches such as smothering, foam application, water-mist systems, etc. could be explored. Testing could first be conducted using small arrays of cells (for example, single boxes of cells rather than full pallets) to assess effectiveness of suppressants in preventing thermal runaway propagation. Full scale testing would be used to validate the most promising approaches.

Gap 3.6: Fire Suppression approaches might be fire stage dependent. There is currently no full scale testing data to determine if variable suppression strategies should be applied depending upon the fire stage. A variety of strategies could be tested using a combination of bench and full scale tests.

Incident Cleanup: Gap 4

Gap 4.1: Appropriate methods for conducting fire overhaul operations should be developed. Such methods should include methods to monitor batteries during breakdown of damaged pallet loads to prevent shorting and high voltage hazards as well as energetic reignition events, recommendations for tools to use in handling debris, and appropriate storage containers for debris. A variety of approaches could be assessed in conjunction with full scale testing efforts.

Gap 4.2: Appropriate procedures should be developed for handling, examining, and disposing of damaged cells and packs after an incident has occurred. These procedures could be assessed in conjunction with full scale testing efforts.

Appendix A
Limitations

At the request of the Fire Protection Research Foundation (FPRF), Exponent has assessed the potential fire hazards associated with lithium-ion batteries. This assessment was intended to be a first step in developing fire protection guidance for the bulk storage and distribution of lithium-ion batteries both alone and in manufactured products. The scope of services performed during this assessment may not adequately address the needs of other users of this report, and any re-use of this report or its findings, conclusions, or recommendations presented herein are at the sole risk of the user.

The opinions and comments formulated in this review are based on observations and information available at the time of writing. The findings presented herein are made to a reasonable degree of engineering certainty. If new data becomes available or there are perceived omissions or misstatements in this report we ask that they be brought to our attention as soon as possible so that we have the opportunity to fully address them.

C. Mikolajczak et al., *Lithium-Ion Batteries Hazard and Use Assessment*,
SpringerBriefs in Fire, DOI: 10.1007/978-1-4614-3486-3,
© Fire Protection Research Foundation 2011